Experiencing Ph

M000115046

Phenomenology is the general study of the structure of experience, from thought and perception, to self-consciousness, bodily awareness, and emotion. It is both a fundamental area of philosophy and a major methodological approach within the human sciences.

Experiencing Phenomenology is an outstanding introduction to phenomenology. Approaching fundamental phenomenological questions from a critical, systematic perspective whilst paying careful attention to classic phenomenological texts, the book possesses a clarity and breadth that will be welcomed by students coming to the subject for the first time.

Accessibly written, each chapter relates classic phenomenological discussions to contemporary issues and debates in philosophy. The following key topics are introduced and explained:

- the methodological foundations of phenomenology
- intentionality as the 'mark of the mental' and the problem of non-existent objects
- perceptual experience, including our awareness of things, properties, and events
- the experience of body, self, and others
- imaginative and emotional experience
- detailed discussions of classical phenomenological texts, including:
    - Brentano's *Psychology from an Empirical Standpoint*
    - Husserl's *Logical Investigations, Cartesian Meditations*, and *On the Phenomenology of the Consciousness of Internal Time*
    - Heidegger's *History of the Concept of Time*, and *Being and Time*
    - Stein's *On the Problem of Empathy*
    - Sartre's *Transcendence of the Ego, Sketch for a Theory of the Emotions*, and *The Imaginary*
    - Merleau-Ponty's *Phenomenology of Perception*.

Also included is a glossary of key terms and suggestions for further reading, making this book an ideal starting point for anyone new to the study of phenomenology, not only in Philosophy but in related disciplines such as Psychology and Sociology.

Joel Smith is Lecturer in Philosophy at the University of Manchester, UK.

Experiencing Phenomenology

An Introduction

Joel Smith

Routledge
Taylor & Francis Group

LONDON AND NEW YORK

First published 2016
by Routledge
2 Park Square, Milton Park, Abingdon, Oxon OX14 4RN

and by Routledge
711 Third Avenue, New York, NY 10017

Routledge is an imprint of the Taylor & Francis Group, an informa business

British Library Cataloguing in Publication Data
A catalogue record for this book is available from the British Library

Library of Congress Cataloging in Publication Data
Names: Smith, Joel Alexander.
Title: Experiencing phenomenology : an introduction / by Joel Smith.
Description: 1 [edition]. | New York : Routledge, 2016. | Includes bibliographical references and index.
Identifiers: LCCN 2015039169| ISBN 9780415718929 (hardback : alk. paper) | ISBN 9780415718936 (pbk. : alk. paper) | ISBN 9781315628639 (e-book)
Subjects: LCSH: Phenomenology.
Classification: LCC B829.5 .S534 2016 | DDC 142/.7–dc23
LC record available at http://lccn.loc.gov/2015039169

ISBN: 978-0-415-71892-9 (hbk)
ISBN: 978-0-415-71893-6 (pbk)
ISBN: 978-1-315-62863-9 (ebk)

Typeset in Times New Roman
by Taylor & Francis Books

Printed and bound in the United States of America by
Edwards Brothers Malloy on sustainably sourced paper

For Ann, Ivy, Sylvia, and Karina

The first elaboration of the field of experience, its phenomena and its givens will offer us an ample supply of difficult and deep problems

Husserl, *Thing and Space*

The term 'phenomenology' expresses a maxim that can be formulated: 'To the things themselves!'

Heidegger, *Being and Time*

Figures

Preface

Phenomenology is one of the most fascinating corners of the philosophical universe. It brings into view something with which we are at once deeply intimate but, at the same time, of which we are typically neglectful: our own experience. Concerned as we ordinarily are with the people and things around us and with the various tasks that fill our days, we seldom stop to consider our experience of this surrounding world. Thinking about phenomenology turns our attention away from the world of things and towards our own experiential life in a way that perfectly exemplifies the knack of philosophical reflection to make the ordinary seem quite extraordinary. It brings home to us the intense curiousness of our everyday experience. My first hope for this book is that it opens up a realm of philosophical questions, making vivid this sense that one's own experience is not something that can simply be taken for granted.

My aim in this book is not to present a standard or established account of phenomenology. In that sense it cannot really be thought of as a textbook. I think of it rather as an advanced introduction to phenomenology. It is an *introduction* in the sense that I assume no prior familiarity with the issues with which I will be concerned, nor am I primarily concerned to advance my own arguments or present my own opinions on the topics discussed. My intention, rather, is to offer a way into the subject. The book is an *advanced* introduction in the sense that it is probably a good idea to have read some other philosophy books before reading this one since it is, in places at least, quite hard. This is, however, unavoidable if we wish to avoid oversimplification.

The book is, in a number of ways, idiosyncratic. I focus on some topics that some would not see as central and ignore some topics that others would consider to be crucially important. Such selectivity is inevitable in a book of any length but, given the intended audience of this one, it is vital to steer a relatively clear path, setting aside some issues for treatment elsewhere, or by others. Whilst I have tried to be even handed, I have no doubt that plenty of people will think my choices misguided. Life is like that.

Since 2007 I have taught an undergraduate course on phenomenology at the University of Manchester and this book has developed out of those lectures. Over the years it has become increasingly clear to me that my students, struggling with difficult and often highly obscure primary texts, were not finding the help they needed in secondary material which seemed to them either itself too hard or otherwise overly simplistic. Furthermore, they often found it difficult to relate what we were studying under the name of 'phenomenology' to other areas of philosophy with which they were already familiar. Phenomenology, so it seemed to some of them, was just a wildly different sort of thing. I hope that reading this book will provide a sense, not just of what phenomenology is, but also of how the work of the classical phenomenologists that I discuss – Brentano, Husserl, Stein, Heidegger, Sartre, Merleau-Ponty – fits into the broader philosophical terrain. The classic phenomenological texts are often written in a style that is inaccessible to those used to the contemporary journal article. But, in my opinion, phenomenology is not radically different from other approaches to philosophy. Rather, the topics discussed in this book are continuous with a range of live issues within contemporary philosophical discourse. My second hope is that this sense of continuity is made manifest in the chapters that follow.

Although I engage with the work of philosophers writing in the first half of the twentieth century, this is not a scholarly book to be filed under the history of philosophy. My main concern, always, is with the *plausibility* of views that can be found in the writings of the above phenomenologists. Nevertheless, it is important, I think, to study the original phenomenological works and, consequently, I do spend a significant amount of time discussing the detail of various classic texts. The chapters can, then, be read as extended commentaries on a selection of these. In my experience, without this level of close engagement, it can be almost impossible to relate an introduction to phenomenology to the often labyrinthine writings that it purports to introduce. In each of the ten chapters, I have focused on one or two primary texts and, at the end of the book, I suggest a small selection of further reading for each topic. In keeping with the introductory goals of the book, I refer to easily available English translations of the original German and French, and have only adjusted those translations where it seemed absolutely necessary (all page references are to the English versions). My choice of texts to discuss has been guided by a combination of philosophical significance and accessibility and I have tried throughout to steer a middle course between the purely historical and the purely systematic, aiming to treat those classic texts with respect but without deference. I'm probably not best placed to judge whether that was a success.

I first encountered phenomenology as an undergraduate while attending Susan James's fantastic lectures on Heidegger. At that time, *Being and Time* seemed to me to be something like a sublime, alien artefact. I was fascinated but didn't really know what to do with it. As a postgraduate, amongst other things, I studied – with Sarah Richmond – Heidegger again and also Sartre's *Being and Nothingness*, during which time a few jigsaw pieces fell into place. My first academic post was at the University of Essex, where I made a concerted effort to understand both Husserl and Merleau-Ponty. It was in this period, and during conversations with my colleagues there, that I developed a real sense of the significance of phenomenology as a whole. It was upon moving to Manchester, however, where for the first time I taught phenomenology, that I really started to gain a concrete appreciation of how the questions animating the Phenomenological tradition relate to issues in epistemology, the philosophy of mind, and the philosophy of language, in which I already had a long-standing interest. It feels as though it has taken me a long time to get to grips with phenomenology, a subject matter that is compelling and frustratingly difficult in equal measure. Hopefully, this book will make that same process a little easier for some of those who are just beginning.

Many thanks to all those teachers, students, and colleagues who have, often unwittingly, contributed to my understanding of phenomenology. At Manchester, in particular, I have benefited from brilliant undergraduates, fantastic teaching assistants, and awesome PhD students. I have also learned a great deal from the writings of a large number of contemporary philosophers working on phenomenological themes. I won't list them here, just take a look at the bibliography. Thanks also to three anonymous referees whose insightful and constructive comments have led to many revisions that I hope have improved the book significantly. Finally, at Routledge thanks to the whole team, especially Helena Power for locating many errors, and Tony Bruce for encouraging me to take on the project and allowing me the time to do, what I hope is, a decent job.

My family have had the greatest impact on the development of this book. Without Ivy, Sylvia, and Karina, it probably would have been written twelve months ago. Without Ann, it would have taken another twelve. I thank them for their patience while I let other jobs go undone so that I could get this book finished. I wrote the book for students, but in a much more important sense it is *for* the four of them.

One last thing: you won't find the terms 'analytic' or 'continental' in the pages that follow. Good riddance.

Joel Smith
Manchester, September 2015

1   The science of experience

we must go back to the 'things themselves'

Husserl, *Logical Investigations*

I sit down at my desk on a sunny September afternoon and open my laptop with the intention of writing this introduction. I glance out of the window and see the sycamore, branches gently swaying in the breeze, its leaves just beginning to show the faintest sign of autumn colour. Noticing its shadow cast across the neighbour's improbably green grass, I picture how the garden would look had I cut the grass yesterday, as planned. My daughter bursts through the door, angry with me for not having taken her swimming this afternoon. Feeling slightly ashamed and conscious of the awkward way in which I have twisted my body in order to face her, I am acutely aware of just how unconvincing my excuse must sound.

This passage, describing a short stretch of experience in which a variety of things appear in a number of different ways, contains all of the major topics covered in this book. The book is about experience. Phenomenology, as the word suggests, is the study of *phenomena*, alternatively *appearances*. This notion of *appearing* is, in turn, related to that of *experience* since things appear in experience. Phenomenology can thus be described as the study of experience and of things *as experienced*. What this amounts to, though, and why one might be interested in such a thing is not obvious. It is, nevertheless, something that captured the imagination of some of the most significant philosophers working in the twentieth century, and continues to do so in the twenty-first.

In the second volume of her autobiography, *The Prime of Life*, Simone de Beauvoir describes how she and Jean-Paul Sartre were introduced to phenomenology:

Raymond Aron was spending a year at the French Institute in Berlin and studying Husserl simultaneously with preparing a historical thesis. When he came to Paris he spoke of Husserl to Sartre. We spent an evening together at the Bec de Gaz in the Rue Montparnasse. We ordered the speciality of the house, apricot cocktails; Aron said, pointing to his glass: 'You see, my dear fellow, if you are a phenomenologist, you can talk about this cocktail and make philosophy out of it!' Sartre turned pale with emotion at this. Here was just the thing he had been longing to achieve for years – to describe objects just as he saw and touched them, and extract philosophy from the process. Aron convinced him that phenomenology exactly fitted with his preoccupations: by-passing the antithesis of idealism and realism, affirming simultaneously both the supremacy of reason and the reality of the visible world as it appears to our senses. On the Boulevard Saint-Michel Sartre purchased Lévinas's book on Husserl, and was so eager to inform himself on the subject that he leafed through the volume as he walked along, without even having cut the pages.

(de Beauvoir, 1960, p. 112)

What was it that so enthralled Sartre? It is the purpose of this chapter to give some sense of what phenomenology is and why it has seemed to many to be such a radical break from the philosophical tradition.

1 Introducing phenomenology

There are a number of ways in which one can get an initial grip on what phenomenology is. Below I outline three: we can compare it with other areas of philosophical inquiry, we can see how phenomenological questions are thrown up in a number of areas, and we can approach it historically. It is also useful to distinguish between broad and narrow conceptions of phenomenology. Since it will be the primary focus of this book, I begin by articulating a narrow conception. What a broader conception amounts to, and the place of the narrow conception within it, will be addressed in §1.3.

1.1 What phenomenology is not

We can see what phenomenology is by way of an understanding of how it compares to other areas of philosophical enquiry such as metaphysics and epistemology. Consider some domain, for example that of concrete, three-dimensional objects such as tables, chairs, protons, and galaxies; that which is sometimes referred to in the epistemological literature as 'the external world'. If we are engaged in metaphysics, we will be interested in the nature of these things. What is it for something to be a concrete, as opposed to an abstract, object? Are all concrete, three-dimensional objects simple or are some composed of smaller parts? Is there a clear distinction between objects and events? And so on.

If we are engaged in epistemology we will be interested in a different range of questions, one centred on the concept of knowledge. What are the different ways we have of coming to know about concrete objects? Do we have any such knowledge at all? If we do, what is the relation between such knowledge and knowledge of other things, for example of our own 'inner' experiences, or of God? And so on.

If we are engaged in phenomenology we will be concerned with a still different range of questions. Unlike the metaphysician, and like the epistemologist, the phenomenologist is concerned not so much with the nature of the external world, but with our mode of access to it. However, for the phenomenologist the central concept is not that of knowledge but is rather that of experience or, equivalently, consciousness. Thus, we might ask: What ways do we have of experiencing concrete, three-dimensional objects? What is involved in the experience of something as concrete rather than abstract? How do concrete, three-dimensional objects *appear* (how are they *presented* or *given*) to us in experience? And so on.

Put in this way, phenomenology may strike some as being closely related to the philosophy of mind. And so it is. Indeed a number of topics currently discussed within the philosophy of mind have a clear phenomenological aspect. One should not conclude from this, however, that phenomenology is just the same thing as the philosophy of mind. The philosophy of mind incorporates questions that are not properly phenomenological, most obviously questions concerning the metaphysics and epistemology of mind. For example, one of the perennial preoccupations of philosophers of mind concerns the right way to think about the relation between consciousness and the brain. Although phenomenology may well have something to contribute to this issue – and to other non-phenomenological debates about the mind – the so called 'mind-body problem' is not an issue that phenomenologists directly address. For the question of the relation between consciousness and the brain is not a question about how things appear *per se*, but rather about the relation between such appearances and what goes on inside our skulls. Phenomenology, it might seem, is radically independent of such a question. For we can ask about how things *appear* without needing to consider how such appearances are realised in the brain.

Philosophers of mind often talk about consciousness, noting that there is 'something it is like' to be in a conscious state. The 'hard problem' of consciousness involves asking how this 'what it is likeness' of consciousness can be located within the physical world. Further, the word 'phenomenology' is often used as a name for this feature. So, for example, bodily sensation is typically said to 'have a phenomenology'. But to say that this is a rather thin characterisation would be a rather monumental understatement. To say that there is something it is like to have a headache is merely the very beginning of a phenomenological account of pain. The phenomenologist quite rightly asks for more.

Phenomenology, understood narrowly, concerns appearance, just as epistemology concerns knowledge. This, I suggest, is the most fundamental way of singling out phenomenology as a distinctive philosophical enterprise. It is this concern to 'describe objects just as he saw and touched them', and to do so in a way that side-steps traditional questions in metaphysics and epistemology, that so excited Sartre. Phenomenology, on this view, is something that can be pursued whilst remaining neutral on perennial metaphysical disputes, turning our back on esoteric debates about the fundamental nature of reality and taking as our theme our ordinary, everyday experience.

1.2 How phenomenological questions arise

Questions of a phenomenological sort arise in both philosophical and scientific contexts. First, consider a philosophical example. The epistemological problem of other minds is sometimes posed in the following way: When I look at you, all that I can see is your behaviour. You are wincing, say. I cannot see the mental state that is causing the behaviour. However, I do not simply judge you to be wincing, I judge you to be in pain. I must, then, be able to reasonably infer that you are in pain from the visible fact that you are wincing. But, for reasons that do not much matter here, there is reason to think that any such inference will not, in fact, be reasonable. Since all I have to go on is your outwardly visible behaviour, I am never justified in believing, and so never know, that you are in pain.

Whether this is a good argument is, for present purposes, neither here nor there. The thing to notice is that the argument makes an implicit phenomenological assumption, one that may or may not be correct. This is that when I look at another person, all that I am visually aware of is their behaviour as opposed to their mental state itself. In asserting this, the proponent of the above argument is taking a stand on the question of how entities of a certain sort – people – are presented in visual experience. But once we notice that this assumption is being made, it becomes less obvious that it is true. What exactly is the most accurate way to characterise the visual experience of another person when they are wincing? Should we say that one's visual experience presents their wincing and their pain is inferred, or should we say that one's visual experience goes, as it were, all the way in to their experience itself? Do we not find it entirely natural to say that we see the pain in someone's grimace, the joy in their smile? As we shall see in Chapter 9, some influential phenomenologists have certainly thought that this is the right way to describe the experience and so would reject the above argument as resting on a false presupposition. This is a clear example of phenomenology making a contribution to non-phenomenological, in this case epistemological, debates about the mind. Furthermore, we see here some of the radical promise of phenomenology since, arguably, getting the phenomenology right has the potential of simply doing away with (at least one version of) the traditional epistemological problem of other minds.

Second, consider an example from psychopathology, one relating to the sort of awareness that each of us has of our own body. People suffering from so-called alien limb syndrome exhibit peculiar behaviour towards one of their limbs, denying that it belongs to them and, in some cases, claiming that it doesn't *feel* as though it is theirs. Since there is also evidence that at least some such patients do continue to feel sensation in the alien limb, this raises the phenomenological questions both of how it is that such patients experience their limb and, more generally, the nature of non-pathological bodily experience. In my everyday experience of my own body, are my limbs and other body parts experienced *as my own*, a feature that the pathological experience perhaps lacks? How is such a 'sense of ownership' related to the fact that I feel sensations located on or within my body? These questions concerning the phenomenology of the body seem pertinent if we wish to give an accurate description of the pathological condition, one that is appropriately sensitive to the complexities of our experiential life. Again, the phenomenological question seems to be fundamental here and yet has the tendency to slip by almost unnoticed. In explicitly asking about the sense of embodiment, phenomenology raises to the status of a philosophical problem something that even those of us engaged in philosophical enquiry ordinarily take for granted.

These considerations give us, I think, an initial sense of what phenomenology is and why it constitutes a significant topic of enquiry. We can see that, at the very least, phenomenology is concerned with offering an accurate description of experience and things as they are experienced. Of course, this is only a thumbnail sketch, and no doubt as many questions have been raised as have been answered. I hope some of these questions will receive answers as we progress.

Another thing that I hope the above examples suggest is that the answers to phenomenological questions are not all obvious. This can seem surprising. There is a long philosophical tradition associated with the view that if anything is absolutely evident to us, it is what our experience is like. Whilst we may be in error as to what there is (metaphysics), or what we know about what there is (epistemology), surely there is no room for doubt about how things *appear* to us (phenomenology)? There is perhaps some truth in this but it is worth noting that there is a difference between something's being evident to those who have a clear focus on the issue, and something's being evident to all no matter how unreflective. Consider, for example, your own visual experience. Looking straight ahead, can you see your nose? It might initially be tempting to answer in the negative, or by saying that you are not sure (after all, the 'edge' of the visual field is a surprisingly difficult thing to describe). But now wiggle your nose. I suspect that something moves within your field of vision. Or close one eye. Now do you see it? If you are anything like me (and I don't think that my nose is unusually long) you can see your nose, but it is pre-sented in what one might describe as a reclusive way. If I focus clearly on the question, it is evident to me that I can see my nose. Before applying such a reflective focus, however, it was not obvious to me at all. You will find this pattern repeated in the chapters that follow. Accurately describing conscious experience, that thing of which we are all most certain, can be a challenging task indeed. Our own experience, despite the fact that we are so familiar with it – perhaps even *because* of the fact that, to use a Heideggerian term, we 'dwell' in it – is a particularly difficult thing on which to gain a firm grip.

1.3 The Phenomenological tradition

In the previous section I have introduced phenomenology in an entirely ahistorical way. However, once one considers the history of the topic one finds a remarkably rich tradition leading from the very beginning of the twentieth century up to the present day (the best overview of this history is Spiegelberg, 1976; also see Moran, 2000). I will use the capitalised 'Phenomenology' to name this historical tradition, reserving the lower-case 'phenomenology' for the subject itself. Another way, then, of introducing phenomenology, is *via* the views of leading figures within the Phenomenological tradition.

Although there is a Phenomenological pre-history, including elements of the work of David Hume (1711–1776), Immanuel Kant (1724–1804), Georg Wilhelm Friedrich Hegel (1770–1831), Franz Brentano (1838–1917), and others, the movement really begins with the work of Edmund Husserl (1859–1938), and continues to some extent, and in a variety of different ways, in the work of Max Scheler (1874–1928), Adolf Reinach (1883–1917), Martin Heidegger (1889–1976), Edith Stein (1891–1942), Roman Ingarden (1893–1970), Jean-Paul Sartre (1905–1980), Emmanuel Levinas (1906–1995), Simone de Beauvoir (1908–1986), Maurice Merleau-Ponty (1908–1961), Paul Ricoeur (1913–2005), Jacques Derrida (1930–2004), and others.

The most important precursor to the Phenomenological tradition is without doubt Brentano's *Psychology from an Empirical Standpoint*, which attempted to place the sciences of the mind on rigorous philosophical foundations. Of particular importance for our purposes is Brentano's introduction of the notion of intentionality (the topic of the next chapter) into modern debates about the mind, and his insistence on the priority of a purely descriptive form of psychological inquiry. Brentano distinguished such descriptive psychology (also called 'psychognosy') from 'genetic psychology' by claiming that since genetic psychology is concerned with 'the conditions under which specific phenomena occur', it cannot proceed without 'mentioning physico-chemical processes and without reference to anatomical structures' (Brentano, 1890–1, p. 4). Descriptive psychology, on the other hand,

> teaches nothing about the causes that give rise to human consciousness [...]. Its aim is nothing other than to provide us with a general conception of the entire realm of human consciousness. It does this by listing fully the basic components out of which everything internally perceived by humans is composed, and by enumerating the ways in which these components can be connected. Psychognosy will therefore, even in its highest state of perfection, never mention a physico-chemical process in any of its doctrines.
>
> (1890–1, p. 4)

As we shall see in this and the next chapter, each of these innovations – the concept of intentionality and the demarcation of a purely descriptive philosophical foundation of the sciences of the mind – had an enormous impact on what Husserl would come to call the phenomenological method.

The work usually considered to constitute the birth of phenomenology is Husserl's *Logical Investigations*. This rich, two-volume work contains, amongst other things, Husserl's celebrated attack on psychologism (the view, to which Husserl himself had previously subscribed, that logic is reducible to psychology), an account of phenomenology as the purely descriptive study of structural features of the varieties of experience, and concrete phenomenological analyses of the notion of meaning, of part–whole relations, and of intentionality. Whilst Husserl's position in *Logical Investigations* does depart in many crucial ways from that of Brentano, it shares – alongside the importance of intentionality and descriptive methods – an adherence to a broadly realist philosophical perspective. Husserl saw his phenomenology as a 'rigorous science' that would provide the basic framework for phenomenological investigations into a wide variety of domains of inquiry, carried out within the spirit of a collaborative investigation (see Husserl, 1910–11). This outlook was shared, at least in broad outline, by a number of the philosophers that this work inspired, most of whom were students or associates of Husserl. Notable examples include, but are certainly not limited to, Scheler's (1913/16) phenomenological ethics, Reinach's (1913) phenomenology of law, Stein's (1917) phenomenological investigation of empthy, and Ingarden's (1931) phenomenology of art.

Husserl continued to refine, modify, and add to his phenomenological analyses throughout the rest of his life, and his output was prodigious. Of particular significance were the first two books of *Ideas Pertaining to a Pure Phenomenology and to a Phenomenological Philosophy, The Crisis of European Sciences and Transcendental Phenomenology,* and *Cartesian Meditations.* Of special interest for the purposes of the present chapter is the account that Husserl gives in these works of the distinction between the phenomenological as opposed to the natural attitude. This is something to which we shall return below. In terms of gaining an understanding of Husserl's overall philosophical development, however, it is crucial to note that these later works represent a shift from his early realist view to an idealist position that, in explicit reference to Kantian transcendental idealism, he termed 'transcendental phenomenology'. In exactly what sense the later Husserl is an idealist, and how his work relates to that of Kant, however, is a matter of much controversy (for differing views see, for example, Philipse, 1995; Smith, 2003, Ch. 4; Zahavi, 2003, Ch. 2; and Woodruff Smith, 2007, Ch. 4).

Whilst Husserl was the Phenomenologist *par excellence*, then, it would certainly be a mistake to suppose that his writings are purely phenomenological in the narrow sense outlined in §1.1, i.e. they address only the structures of experience considered in isolation from their metaphysical and epistemic significance. Husserl's work is, rather, shot through with detailed discussion of the broader metaphysical and epistemic context in which he sees his phenomenological analyses as fitting. In addition to the relation between Husserl's phenomenology and Kantian transcendental philosophy, examples would include his detailed criticism of naturalistic epistemology, his account of the relation between fact and essence, and his regional ontology more generally. We can, then, distinguish the narrow conception of phenomenology with a broader conception, one that incorporates this wider perspective.

If this is true of Husserl, it is doubly so of the majority of those that follow him in the Phenomenological tradition. A case in point, and an absolutely pivotal figure in the development of Phenomenology, is Heidegger. As a junior colleague of Husserl's, Heidegger was seen by Husserl and others as the rightful bearer of the Phenomenological torch, set to continue Husserl's work of setting the groundwork of philosophy as a rigorous science. It soon became clear, however, that this was not to be so. Whilst Heidegger's early work is deeply influenced by Husserlian phenomenology, it took Phenomenology in a significantly new direction, as is evident from his master work, *Being and Time*, and also the lecture courses *History of the Concept of Time*, and *The Basic Problems of Phenomenology*. Indeed, reading *Being and Time* is an experience quite unlike reading Husserl – I am tempted to say quite unlike any other. Combining aspects of Husserlian phenomenology with themes from the Aristotelian tradition – primarily a relentless pursuit of the question of the meaning of being – Heidegger referred to his own phenomenological investigations as 'fundamental ontology'. His early work combines a focus on distinctively phenomenological questions with a penetrating critique of Husserl, and an account of the structures and meaning, *for us*, of human existence, that has had a long lasting influence, including on the existentialist phenomenology that was soon to flourish in France. Whilst commentators are divided on the question of the relation between Husserlian and Heideggerian phenomenology (see, for example, Dreyfus, 1991; Blattner, 1999a, Introduction; Crowell, 2001, Part 2; Crowell, 2004), and his late work is not obviously to be thought of as phenomenological at all, there is no doubt that he is, after Husserl, the central figure in the evolution of the Phenomenological tradition.

Phenomenology made its way to France, in large part, by way of Levinas's (1930) book on Husserl. Each of Sartre, de Beauvoir, and Merleau-Ponty combined an interest in Husserlian phenomenology, and in particular a Heideggerian emphasis on the primacy of active experience, with a broader concern with psychological, sociological, and political issues. Sartre's early works, *The Imagination, The Transcendence of the Ego, Sketch for a Theory of the Emotions*, and *The Imaginary*, offer both original accounts of their respective subject matters and also sustained investigations as to the limits of Husserlian phenomenology. It is in his *Being and Nothingness*, and de Beauvoir's *The Ethics of Ambiguity*, however, that Heidegger's influence can be most clearly seen, with the two existential phenomenologists taking their lead from Heidegger's account of authentic existence and developing phenomenological accounts of consciousness, freedom, and concrete human relations that perhaps define the term 'existentialism'. Merleau-Ponty, before his falling out with Sartre over political issues, worked alongside Sartre and de Beauvoir, also combining elements of Husserlian and Heideggerian phenomenology. More than any other phenomenologist before him, Merleau-Ponty brought the tradition of Phenomenology into contact with psychological theories, an undertaking that is, arguably, a precursor to the contemporary project of naturalising phenomenology (Petitot et al., 2000; although see Gardner, 2015 for a different take on Merleau-Ponty's project). Taken together, the French Phenomenologists outlined a distinctive approach to philosophical questions that pursues the project of Phenomenology in new and highly innovative ways.

In the work of any one of the philosophers working within the Phenomenological tradition, we see a complex weave of phenomenological, metaphysical, and existential concerns. That is, each pursues a project that can be seen as phenomenological in the broad sense indicated above. Separating out the *narrowly* phenomenological threads can sometimes be a difficult task. Nevertheless, this is an introduction to phenomenology narrowly construed and so it is one that I attempt. This is not innocent since some will object that phenomenology in the narrow sense cannot be extracted from its broader philosophical context without causing irreparable damage. I hope to show that this isn't true, and that something of value can be drawn out in this way. It is true, however, that at times the line between narrowly phenomenological concerns and bordering issues concerning, for example, ontology becomes blurred. As such, the book occasionally dips a toe in such deep waters.

Whilst no member of the post-Husserlian Phenomenological tradition is a straightforwardly Husserlian phenomenologist, in each case Husserl sets the phenomenological agenda. This remains the case, with a great deal of the contemporary interest in both phenomenological methodology and phenomenological topics drawing inspiration from Husserl's work. Accordingly, Husserl's views are the touchstone and it is to his account of the distinctive features of phenomenological method and subject matter that we now turn.

2 The science of experience

As conceived by Husserl, phenomenology is not a theory or a body of philosophical claims – it is not a doctrine. Rather, phenomenology is a set of philosophical questions and an approach towards them – it is a method. But what is the method? Two notions are central: the phenomenological reduction (or *epoché*) and the eidetic reduction. The role of the phenomenological reduction is to enable accurate description by reflectively focusing the attention on *experience* rather than the things experienced. The role of the eidetic reduction is to focus on the *essential* features of experience. Husserl saw these two as vital to the practice of phenomenology. Once we see why, we will have a grip on the sense of Husserl's phenomenological slogan, 'Back to the things themselves!'

Husserl presented his views on phenomenological method many times and, as such, there is no clearly canonical text in this regard. However, of particular use is his 1917 inaugural lecture at Freiburg, 'Pure Phenomenology, its Method, and its Field of Investigation', an excellent text from which to glean a first understanding of what Husserlian phenomenology is all about. We can begin with Husserl's initial characterisation of phenomenology as the 'science of consciousness',

> a science of objective phenomena of every kind, the science of every kind of object, an 'object' being taken purely as something having just those determinations with which it presents itself in consciousness and in just those changing modes through which it so presents itself. It would be the task of phenomenology [...] to investigate how something perceived, something remembered, something phantasied, something pictorially represented, something symbolized looks as such.
>
> (1917, pp. 126–7)

This short passage raises a number of questions: What does Husserl mean by the terms 'phenomena' and 'object'? What does he mean by saying that phenomenology investigates how the objects of various sorts of experience 'look as such'? In what sense does Husserl think of phenomenology as a science? To answer these questions, we must turn to the two cornerstones of Husserl's phenomenological method: the phenomenological and eidetic reductions.

2.1 The phenomenological reduction

In our everyday lives we occupy what Husserl calls the 'natural attitude'. In the natural attitude, our attention is occupied by the objects and events around us, the objects of which we are conscious. Furthermore, we take it for granted that the world around us exists independently of us and our awareness of it. The tree that I see through my window, the desk at which I am working, the song I am listening to, the friends and family that I may bring to mind, these one and all are real and are so whatever my own view of them. These are what Husserl calls 'Objects in the pregnant sense of the word' (1917, p. 127), mind-independent things whose reality transcends – goes beyond – our experience of them. Objects, in this sense, can be contrasted with the broader category of objects, which are simply those entities of which we can be aware regardless of whether or not those things exist in reality, are mind-independent, and so on. Thus, whilst tables and chairs are Objects, the traditional notion of a sense-datum, being mind-dependent, is merely an object. The broader category of objects will be the focus of Chapter 2, and we will come across sense data in Chapter 3.

This belief in the reality of the world implicitly informs and permeates our everyday life. Without it, we wouldn't know which way to turn. We would have no faith in the solidity of the world around us. Nevertheless, the phenomenological method requires us to leave the natural attitude, adopting instead a reflective attitude in which we attend not to the objects of our experience but to our experience of objects. In the reflective attitude I focus, for example, not on the tree beyond my window, but on the awareness that I have of it. In addition the phenomenological reduction requires us to 'place in brackets', or 'put out of action', our belief in the independent existence of the objects of experience, i.e. we bracket Objects. Thus, not only do I focus on my experience of the tree, rather than the tree itself, I bracket the fact that I take it to be a real tree and not a mirage. This is the phenomenological reduction, or *epoché*.

What is bracketed in this way is everything that it is possible to doubt. Anyone acquainted with Descartes' *Meditations on First Philosophy* will know that this is a great deal indeed, including all worldly Objects around one, even one's own body. As Husserl puts it, 'what is experienced through external experience always allows the possibility that it may prove to be an illusory Object in the course of further experiences' (1917, p. 128).

The connection to Descartes, one about which Husserl is explicit, is not so close that the phenomenologist is taken to be in the same position as Descartes' thinker, left in doubt as to whether the external world really exists. The phenomenologist does not cease to believe in the surrounding world, but rather makes no use of it in phenomenological investigations. What this means is that when we engage in phenomenology, we rely on nothing that presupposes the actuality of the world around us. According to Husserl, this includes all empirical sciences since, on his view, these are concerned with Objects; with things the reality of which transcends our awareness of them. This is true not only of physics and chemistry, which are concerned with the constitution of mind-independent nature, but also of biology and even psychology, which are concerned with organisms and cognitive systems respectively. These scientific disciplines presuppose that the entities with which they are concerned exist in reality. As such, on Husserl's picture, they and their results, as valuable as they are in other contexts, do not survive the phenomenological reduction and so have no place in phenomenology.

On Husserl's view, the reduction allows us to describe that which is given in experience purely *as it is given*, in a way entirely unswayed by the realistic prejudices of the natural attitude and empirical science. To get a sense of what Husserl has in mind here consider, again, the case of how others are presented to us in perceptual experience. The sceptical argument that I earlier sketched presupposed that, in perceiving another, we are aware merely of their outward behaviour and, so, any judgement that they are in some mental state must result from an inference from such behaviour. I suggested that this pre-supposition is questionable and that it ought to be subjected to scrutiny. Perhaps others are perceptually given not just as behaving but also as possessing various mental states. The phenomenological reduction gives us a way to approach this question of how others are given. Suppose, however, that one approaches it from within the natural attitude. Then, one might be influenced by the thought that, since mental states are identical to, or supervene on, states of the brain and central nervous system, and since the brain and central nervous system are, ordinarily at least, hidden from view, surely *one could not possibly* see another's mental states. It *must* be that experience presents mere behaviour. In this way an empirically grounded theory pushes us to answer the phenomenological question in a particular way. But, in doing so, it leads us away from a consideration of the experience itself. Rather than reflect on what the experience of another is *actually* like, we form a view about what it *must be* like. This, thinks Husserl, is to approach phenomenology in exactly the wrong way.

Within the confines of the phenomenological reduction, with the world of Objects bracketed, we reflect our attention towards our experience itself. This Husserl calls 'pure reflection' (1917, p. 129). In pure reflection – distinguished from reflection within the natural attitude – we attend to all and only those objects that survive the phenomenological reduction. These *phenomena* are to be the subject matter of phenomenology.

If phenomena are those and only those things that survive the phenomenological reduction, and the reduction is to be applied to anything whose existence may be doubted, then phenomena must be all and only those things with indubitable existence. But what, we may ask, passes this test? Husserl maintains that phenomena include both conscious experiences and their objects as they appear in experience. He writes that the concept *phenomenon* covers 'the whole realm of consciousness with *all* of the ways of being conscious of something and all the constituents that can be shown immanently to belong to them', including the objects of such conscious experiences, 'just the way they are given to consciousness' (1917, p. 126). The idea here is that whilst, as I look through my window, I may doubt the reality of the tree that I seem to see, I may not doubt either the reality of the visual experience itself (the way of being conscious of something) or the appearance of a tree (the object just the way it is given in experience).

Consider first the appearance of the tree. It is a familiar thought that whilst my senses may mislead me as to what objects exist, they cannot do so with regard to what objects *seem* to exist. That there is a tree beyond my window is open to question and can be verified, at least to the satisfaction of everyday standards, by walking around it, touching it, and so on. It is conceivable that such tests will reveal that there is, in truth, no tree; that it was a trick of the light. But that there appears to me to be a tree beyond my window is neither open to question nor verification in this way. As Husserl puts it, this object – the apparent tree – is given *absolutely* (1917, pp. 127–8). That is, there is nothing more to learn about it, no obscured sides or hidden depths, beyond what is already available to me in enjoying the conscious experience in which there appears to be a tree beyond the window. If we limit our description of the apparent tree to just those features the tree appears to have then it would seem that appearance and reality coincide. It is this *absolute givenness* that puts the apparent tree beyond the reach of sceptical doubt. It is crucial, at this point, to emphasise that, according to Husserl, nothing is here being assumed about the relation between appearance and reality. For example, there is no implicit suggestion in the above that the real tree is somehow hidden 'behind' the apparent tree. Although we will meet such a view later in various guises (e.g. the sense-data theory of perceptual experience), mere talk of appearance should not be taken to be committed to so controversial a position.

Much of the above applies to the visual experience in which the tree appears. Of course, given that we can only, literally, see things that exist, the revelation that there is no tree would also undermine the claim that I see a tree. But it would not undermine the claim that I *seem* to see a tree, alternatively put that I am enjoying a visual experience as if there were a tree before me. Again, even the radical supposition that I have no body, and so no eyes with which to see, could not cast doubt on the reality of my visual experience itself. Once more, this is associated with the thought that whilst the objects of visual experience are presented to me only partially, with obscured sides and hidden depths, visual experience itself is given absolutely. All there is to the nature of the visual experience itself is available to me 'from within' simply in virtue of my having it.

We are now in a position to answer the first two of our three questions concerning Husserl's description of phenomenology as the 'science of consciousness'. These were: What does Husserl mean by the terms 'phenomena' and 'object'? And, What does Husserl mean by saying that phenomenology investigates how the objects of various sorts of experience 'look as such'? In answer to the first, we may say that phenomena are those objects that are absolutely given, and so indubitable, and so survive the phenomenological reduction. It is these that form the subject matter of phenomenology. In answer to the second we may say that the phenomenologist is concerned to describe both experiences and their objects just as they appear, or how they 'look as such'. To go beyond appearances would be to go beyond that which is absolutely given, and so stray beyond the proper boundary of phenomenology.

2.2 The phenomenological reduction on trial

The phenomenological reduction has had a chequered history within the Phenomenological tradition with many rejecting the letter, and even the spirit, of the reduction as Husserl describes it. In the preface to his *Phenomenology of Perception*, for example, Merleau-Ponty wrote that '[t]he most important lesson of the reduction is the impossibility of a complete reduction' (1945, p. lxxvii), the suggestion being that the commonsense realism of the natural attitude is, in some respect, inescapable. Merleau-Ponty's view of the phenomenological reduction is controversial, complex and, at least in part, bound up with his attitude towards the transcendental elements of Husserlian phenomenology. This is not the place to examine those issues, taking us, as they do, well into Merleau-Ponty's philosophy (elements of which will be discussed in later chapters) and into the territory of the transcendental and existential context in which much phenomenological work has been carried out. Nevertheless, there are some reasonably clear objections to the phenomenological reduction that resemble Merleau-Ponty's to the extent that they stem from the commonsense realism of the natural attitude itself.

The phenomenological reduction will, no doubt, strike some as deeply misguided. Anyone with even a passing familiarity with neuroscientific or cognitive psychological work on, for example, vision will be well aware that there is a great deal more to visual experience than can be known simply in virtue of attending to visual experiences 'from within'. No account of vision could possibly be complete without, at the very least, providing a detailed explanation of the sensitivity of the eye to light and of the workings of the visual areas of the brain. The first objection, then, is to accuse the phenomenologist of wilfully disregarding everything that we know about the relation between visual experience, on the one hand, and neural function and information processing, on the other.

This is a serious objection and, to see how a phenomenologist might respond to it, let us temporarily put vision to one side and consider instead the example of pain. There is a great deal of intuitive plausibility to the thought that for pain, appearance and reality coincide. That is, if one has an experience that seems in every way like an experience of pain, then one is having a pain experience. This can be contrasted with the above example of seeing a tree. In that case, one can have an experience that seems in every way like seeing a tree, yet not be seeing a tree. This is for the reason that if there is no tree there, I can't be seeing one, although it might seem to me as if I am. Now, consider a creature having an experience that is, for it, just like our pain experiences are for us. It seems that we would judge this creature to be in pain, whether or not it had the same neural properties as we do when we are in pain. This is a point familiar from discussions in the philosophy of mind where it is often claimed that consciousness is 'multiply realisable': we intuitively accept that *at least in principle* a creature could be in pain even if it had a brain very different from our own (Putnam, 1967).

Whether this intuitive thought about pain is in fact true and, if it is, what follows from it about the nature of pain, are both controversial. Nevertheless, it certainly seems far too quick to suppose that our scientific knowledge of pain will be in any way threatened by these observations. Rather, the very most that could be concluded is that neither neuroscience nor cognitive psychology tell us what the *nature* or *essence* of pain is. Rather, what they tell us is how pain is, as it is said, 'realised' in creatures like us. Further, everything that has just been said about pain can also be said about visual experience or, seemingly, conscious experience of any sort at all. So, whilst it is evident that an account of the workings of eye and brain are required to explain human visual experience, the fact that *in principle* a creature with a very different physiological make-up could enjoy an experience of the same sort suggests that such an account would not tell us about the nature of visual experience *per se* but rather about how visual experience is realised in humans. Further, as long as we are careful to focus only on phenomena – on, in the case of visual experience, apparent seeing – then it can indeed seem plausible to suppose that, as was suggested above, the nature of this experience can be gleaned from simply attending to the experiences themselves. For, whilst seeing a tree requires there to be a tree that one sees, having a visual experience in which there appears to be a tree before one seemingly does not. After all, we can easily imagine a hallucination of a tree, or Descartes' familiar thought experiment in which we are deceived by an evil demon into thinking that there is a world at all. Thus, on this way of thinking, it can seem that the nature of visual, or any other, experience can be determined by means of a reflection on one's own experience. Indeed, it can seem plausible to suppose that experience can *only* be understood in this way.

Husserl makes just this point when he distinguishes between 'pure phenomenology' and 'psychological experiencing' (1917, p. 129). His point is that any account of experience and the different forms that it can take that seeks to locate it within the realm of Objects (and here we can think of both psychology and much of contemporary philosophy of mind), will not succeed in articulating the very nature or essence of conscious experience. For any claim about how, to return to the example of vision, visual experience relies on information processing in the visual cortex, will be *contingent*. We can easily imagine creatures with visual experience yet without a visual cortex. The *essence* of visual experience must be understood in some other way. The claim distinctive of Husserlian phenomenology is that this other way is, of necessity, *via* pure reflection from within the phenomenological reduction, in which '[c]onsciousness is taken purely as it intrinsically is with its own intrinsic constituents, and no being that transcends consciousness is coposited' (1917, p. 129). On the Husserlian picture, then, employing the phenomenological reduction does not mean illegitimately ignoring the sciences of the mind. Rather, it enables us to focus on the essential features of experience, something that scientific disciplines are incapable of doing. We will return to the notion of essence in the next section.

A second reason to be sceptical of some of the claims made in the name of the phenomenological reduction rests on what is sometimes called the 'naïve realist' view of conscious experience. The issue is most pressing, perhaps, in the case of visual perception. When I see the tree beyond my window it is difficult to describe my experience without mentioning that it is a tree that I see. That is, in characterising visual experience we appeal to Objects, to things the reality of which transcends our experience of them. Naïve realist views of experience hold that in experiencing a tree I stand in a specific relation of acquaintance to that particular tree. Furthermore, the fact that I stand in this relation constitutes my experience's being as it is; it constitutes there appearing to be a tree beyond the window. As Brewer describes the view, 'perceptual presentation irreducibly consists in conscious acquaintance with mind-independent physical objects' (Brewer, 2011, p. 94; also see Crane, 2006b).

If true, this position undermines the claims made on behalf of the phenomenological reduction. On the naïve realist view, we should not agree that the primary task is to provide an account of the essence of visual experience, considered independently of whether the objects of that experience actually exist, i.e. are Objects. For, according to the naïve realist view, whilst experiences of trees essentially depend on the existence of trees, experiences that appear to be, but are not in fact, experiences of trees, do not. That is, whilst, 'the fundamental nature of perceptual experience is to be given precisely by citing and/ or describing those very mind-independent physical objects of acquaintance' (Brewer, 2011, p. 94), the same cannot be true of a hallucination in which there are no such Objects with which we are acquainted. On the naïve realist view, then, when we see a tree we are in a very different type of experiential state than when we hallucinate a tree. One is, essentially, a way of being acquainted with Objects, the other only seems to be.

According to the naïve realist, then, the phenomenological reduction would not provide a reliable route to determining the nature of visual experience. For in bracketing the existence of Objects, it fails to respect the difference between seeing a tree and enjoying a perfect hallucination of a tree. The phenomenological reduction would, rather, represent a version of what McDowell has called the *highest common factor* view of experience, according to which just the same 'is available to experience in deceptive and non-deceptive cases alike' (1982, p. 386). This is inconsistent with naïve realism and so, far from being the neutral starting point that Husserl suggests, embodies substantive and controversial philosophical presuppositions.

Naïve realism is not the majority position, and the highest common factor view can certainly seem overwhelmingly intuitive for the reason that it neatly explains why it is that an experience in which I really am aware of a transcendent Object can be subjectively indistinguishable from one in which I only appear to be. According to a highest common factor view, they are subjectively indistinguishable because they are experiences of exactly the same sort. The naïve realist must deny this, offering another explanation of the apparent indistinguishability of seeing and hallucinating (for discussion, see Fish, 2009, Ch. 4). Nevertheless, Husserl thought of phenomenology as a *presuppositionless* science of experience, writing that 'phenomenology, by virtue of its essence, must claim to be "first philosophy" [...] therefore it demands the most perfect freedom from presuppositions' (Husserl, 1913, p. 148). The phenomenological reduction and its bracketing of Objects is the tool with which Husserl attempted to achieve this freedom from presuppositions. However, the very existence of the naïve realist view with its apparent inconsistency with the use to which the Husserlian phenomenologist puts the phenomenological reduction, suggests that this cannot be true. The idea that the phenomenological reduction opens up the possibility of determining the essence of the varieties of experience itself relies on its own presuppositions. Naïve realism, then, requires a response from anyone espousing the phenomenological reduction as a cornerstone of the phenomenological method.

One option here is to simply accept that phenomenology is not pre-suppositionless in the way that Husserl claims. In fact, this may be motivated on a number of fronts. For example, not only is there a question about naïve realism but, as we also saw above, Husserlian phenomenology makes claims about essence and so, insofar as it is committed to the existence of essences, lacks neutrality on that score as well. The rejection of presuppositionless is a move that Heidegger explicitly makes, writing that '[t]he interpretation of something as something is essentially grounded in fore-having, fore-sight, and fore-conception. Interpretation is never a presuppositionless grasping of something previously given.' (1927a, p. 146; also see §63). Heidegger's claim here, which will be relevant again below, is that presuppositionlessness is a myth. All interpretation of experience and of things as experienced depends on a prior conception of them. Of course, simply recognising this point does not vindicate the employment of the phenomenological reduction, as it does nothing to show that naïve realism is false. Abandoning the goal of pre-suppositionlessness may, however, be a necessary first step in such a vindica-tion. I will have more to say about the naïve realist view in Chapter 3.

2.3 The eidetic reduction

It is often said that phenomenology is a descriptive enterprise, its central task being to provide a clear, undistorted *description* of the ways in which things appear. Husserl himself speaks of the 'observations, descriptions, theoretical investigations' (1917, p. 128) that are characteristic of phenomenology. This can be distinguished from the project of giving, for example, causal or evolu-tionary explanations of Objects, which is the job of the various empirical sciences. This distinction is clear enough. Consider, once again, the case of pain. We might offer an account of pain in terms of its proximal causes, such as the stimulation of nociceptors. Or again, we might present an account in terms of the function of pain, for instance the fact that it motivates us to avoid bodily damage. But each of these projects presupposes that we have a prior grip on what pain is. We need to know for what we are attempting to find the causes or function. Explanations of pain rely, that is, on a *description* of pain, for example as an unpleasant sensation located in or on the body. The task of phenomenology, as Husserl conceives it, is to offer such descriptions in a detailed and systematic fashion.

For such descriptions, Husserl has high hopes, writing that '[w]hat phenomenology wants, in all these investigations, is to establish what admits of being stated with the universal validity of theory' (1917, p. 127), and that 'pure phenomenology proposes to investigate *the realm of pure consciousness and its phenomena* not as de facto existents but as pure possibilities with their pure laws' (1917, p. 132). As I suggested in the previous section, Husserlian phenomenology has as its goal an appreciation of the essential features of the varieties of experience. Its descriptions are to be descriptions of the 'pure laws' of experience. The phenomenologist is not concerned to describe the contingent features of, for example, pain, which would have only local significance, but those features that admit of 'universal validity'.

It is easy to see why one might be sceptical of such a grand ambition. The Husserlian phenomenologist seeks to describe phenomena by first bracketing all transcendent Objects, then reflectively attending to the ways things appear in their own conscious experience. But is this not a recipe for an entirely subjective, idiosyncratic, and untestable mishmash of claims about *how things seem to me*? And how is Husserl's ambition to the 'universal validity' of 'pure laws' to be squared with what seems to be the irredeemably subjective starting point of the phenomenological reduction? Put in other terms, how can Husserl justify his claim that through phenomenology we can come to know the *essence* or *nature* of phenomena?

Related to this concern is another concerning Husserl's contention that phenomenology is a scientific enterprise, 'the science of *pure* consciousness' (1917, p. 129). Recall the third of our questions concerning phenomenological method: in what sense does Husserl think of phenomenology as a science? Since the phenomenological reduction has, by Husserl's own lights, set aside all empirical science of the mind, it is difficult to see how he can substantiate his claim that phenomenology 'is inferior in methodological rigor to none of the modern sciences' (1917, p. 124).

Husserl's answer to these twin concerns lies in a second, eidetic, reduction. The eidetic reduction involves bracketing all contingent matters, instead attending only to the essential features of experience and things as they are given in experience. Husserl offers an analogy with pure mathematics and pure geometry which, like pure phenomenology, are not concerned with the actual existence of Objects (1917, pp. 130–1). Just as in pure geometry the properties of, say, Euclidean space can be studied without any commitment to the actual existence of Euclidean space, so in pure phenomenology the properties of phenomena can be studied without any commitment to the actual existence of corresponding Objects. Phenomenology, on Husserl's understanding, is scientific in the same sense as is mathematics. It is not a natural science, following tried and trusted empirical methods, but is an *a priori* science of essence. The bracketing of Objects, then, does not entail that phenomenology cannot be scientific in this broader sense (note that the German *Wissenschaft* has a somewhat broader application than does the English 'science', incorporating any rigorous or systematic research discipline).

But this still does not show how the charge of subjectivism can be deflected. We want to know how it is that phenomenology can be like mathematics, how the phenomenologist can come to knowledge of essence. On Husserl's view, to answer this we must turn to our capacity for free variation in the imagination. The pure geometer, he says,

> is not bound to shapes observed in actual experience but instead inquires into possible shapes and their possible transformations, constructing *ad libitum* in pure geometric phantasy, and establishing their essential laws, in *precisely* the same way pure phenomenology proposes to investigate *the realm of pure consciousness and its phenomena.*
>
> (1917, p. 132)

The clue here is the connection between imagination ('phantasy') and possibility. The thought is that, at least when it comes to phenomena, our capacity to imagine something is a good guide to its possibility. If something can be imagined, that is a reason to think that it is possible; if it cannot, that is a reason to think that it is not. To discover, therefore, the essence of some phenomenon, we imagine variations on it to determine what is and is not a possibility for it. I imagine an object of a particular kind and vary it in some way. If I cannot so vary it without its ceasing to be an entity of that kind, then that initial way must be essential to that kind, part of its essence. Consider, for example, the tree that I see beyond my window. In imagination I may freely vary its colour, shape, or apparent spatial location without thereby imagining something other than a tree appearing to me. However, if I attempt to imagine it as occupying no apparent spatial location at all, then I have ceased to imagine a possible object of visual experience at all. From this, we may conclude that appearing spatially located is part of what it is to be an object of vision, part of its essence. In this way, we may arrive at knowledge of the essence of experience and its phenomena, thereby laying the charge of subjectivism to rest (see Husserl, 1913, §70).

This method is not entirely different from the standard method of conceptual analysis: imaginative thought experiments (cf. Thomasson, 2007). Consider the well-known Gettier (1963) examples widely considered to show that knowledge is not true, justified belief. We begin with a hypothesis (that knowledge is true, justified belief), we imagine a scenario about which we are invited to judge that there is true, justified belief without knowledge, and we conclude that the hypothesis is false. It may well seem, then, that to the extent that thought experiments provide an acceptable philosophical methodology, Husserl's eidetic reduction and its method of free variation in the imagination is on safe ground.

2.4 The eidetic reduction on trial

Things may not be so simple, however. An immediate worry about Husserl's eidetic method is that it is circular (Bell, 1990, p. 195). In order, it may be claimed, to determine whether some imagined object is an instance of some particular kind of thing, we already need to have some, perhaps implicit, conception of the nature of that kind. For example, if I judge that an imagined non-located entity cannot be an apparent tree, I must already have some grasp of what it is for a tree to appear to me. Otherwise, if I ask whether something is an apparent tree, I would not be in a position to answer. Given this, our intuitive judgements about imagined scenarios can at best serve to articulate knowledge we already possess. They can teach us nothing new.

It may be that this is less an objection to Husserl than an expression of his position. The Husserlian phenomenologist is not, after all, attempting to explain how we acquire the concepts in question. The point, rather, is to gain a clear, explicit understanding of the nature of the varieties of experience and objects as experienced. It is not surprising, for example, that to gain a clear understanding of the nature of, say, emotion one must already possess an implicit understanding of what does and what does not count as emotion. As Bell puts it, on this view the eidetic reduction would be 'aimed at making clearer or more distinct the universal "ideas" which we already possess and use in everyday life, albeit in a more or less indistinct and unclear way' (1990, p. 195). This is an issue to which we will return in the discussion of Heidegger's attitude towards Husserlian phenomenology.

A second concern relates to the purported role of imagination in teaching us of the essence of phenomena, of what is and is not a possible instance of a given kind. That there is a close connection between imagination and possibility is an influential thought, indeed Hume went so far as to say that

> 'Tis an established maxim in metaphysics, *That whatever the mind clearly conceives includes the idea of possible existence*, or in other words, *that nothing we imagine is absolutely impossible.*
>
> (1739–40, p. 32)

There is room for doubt, however. First, a very general worry. While there seems to be a reasonably clear explanation of why it is that perceptual experience discloses facts about the way the world actually is, there is no obvious explanation of why it should be that our faculty of imagination would similarly disclose facts about the ways it is possible for the world to be. Plausibly, my visual experience of the tree is causally related to the tree's reflecting light in my direction. Presumably no analogous causal explanation is available in the case of our knowledge of merely possible things. As such, the eidetic method relies on a connection that is at best somewhat mysterious.

It may be responded that however mysterious the general connection between imagination and possibility, there is an available explanation in the specific case of phenomena. Husserl need not suppose that, *via* the method of free variation in the imagination, we can come to knowledge of possibilities regarding transcendent Objects. He is concerned, rather, with phenomena:

entities constitutively connected to conscious experience itself. This gives us some reason to suppose that the imagination, a form of experience, may be well placed to do the work required of it. To make this stark, consider the phenomenology of imagination. If we are concerned to discover the essence of imaginative phenomena (of things insofar as they appear to the imagination), we would do well to see what we can imagine! Using the imagination is surely a candidate for the best available guide to what is imaginable, i.e. how an imagined object must appear to the imaginer. Things are less straightforward, of course, with modes of experience other than the imagination. However, insofar as we suppose that there is a close connection between the imagination and those other modes – between visualisation and vision, for example – we may have the beginnings of a defence of the view that there is no mystery here to solve. We will return to the imagination, and to this question of the relation between imagination and other modes of experience, in Chapter 6.

3 Back to the things themselves!

3.1 Husserlian phenomenology

Discussions of phenomenology sometimes tend to get bogged down in methodological issues. This was certainly a problem to which Husserl was susceptible. Nevertheless, some understanding of the method of Husserlian phenomenology is crucial for an appreciation not only of the Phenomenological tradition but of the concrete phenomenological analyses with which this book is primarily concerned. With the above, albeit brief and incomplete, discussion of Husserl's 'science of pure consciousness', we are now in a position to appreciate the significance of his slogan, 'to the things themselves'. This is neatly introduced in the Introduction to *Logical Investigations*:

> we can absolutely not rest content with 'mere words' [...]. Meanings inspired only by remote, confused, inauthentic intuitions – if by any intuitions at all – are not enough: we must go back to the 'things themselves'.
>
> (1900–1, Vol. 1, p. 168)

In this particular context, Husserl has logic in mind but the point is perhaps more obviously made with respect to those disciplines that deal, either directly or indirectly, with experience itself. Psychology, and the rest, thinks Husserl, must begin with a clear understanding of their own central concepts. But this grasp is ultimately given only through a phenomenology that adopts both the phenomenological and eidetic reduction. It is this enterprise that enables us to achieve explicit knowledge of the nature of experience and its phenomena. Without such a phenomenological grounding, all we have are 'mere words' without a clearly perceived meaning, floating free of any basis in experience. Such an understanding must be grounded, ultimately, in the description of phenomenologically and eidetically reduced experience (see Husserl's 'Principle of all Principles', in §24 of *Ideas I*). Husserlian phenomenology, then, stands between empiricism and rationalism. Like the rationalist the phenomenologist endorses *a priori* knowledge; like the empiricist the phenomenologist seeks to ground all knowledge in experience.

Thus, phenomenology is in this sense prior to psychology. In Husserl's words:

> [p]ure phenomenology's tremendous significance for any concrete grounding of *psychology* is clear from the very beginning. If all consciousness is subject to essential laws, then these essential laws will be of most fertile significance in investigating facts of the conscious life of human and brute animals.
>
> (1917, p. 132)

Not only, then, is phenomenology distinct from psychology, it is more fundamental than it for an understanding of experience. On Husserl's view, phenomenology is the method of gaining the clear understanding of the nature of conscious phenomena required for a well grounded empirical psychology. As suggested above, empirical research into, say, pain requires a clear understanding of the nature of pain and this, in turn, requires phenomenology. Exactly similar points might be made about certain ways of approaching the philosophy of mind. After all, philosophers concerned with the question of whether pain is identical to a neural state would surely benefit from a clearly articulated understanding of the nature of pain! A psychological or philosophical theory may be ingenious, subtle, and elaborately constructed, but without a proper phenomenological grounding we will lack a clear understanding of the concepts it employs. As Sartre rhetorically asks, 'before experimenting, isn't it appropriate to know as exactly as possible *on what* one is going to experiment?' (1936, p. 127).

But how does phenomenology itself avoid the charge that it deals with 'mere words'? How is it more closely related to 'intuition' – i.e. particular experience – than are either psychology or non-phenomenological approaches to the philosophy of mind? How does it manage to go back to the 'things themselves'?

The answer to this lies in Husserl's doctrine of eidetic intuition, an aspect of Husserl's position that can seem hard to swallow. Recall that, according to Husserl, phenomenology is descriptive not explanatory. Thus, we would expect him to claim that the knowledge of essence that is gained *via* the eidetic reduction is attained by describing experience, experience of essences. And this is exactly what he does tell us. According to Husserl, knowledge of the nature of experience is ultimately grounded in a non-inferential 'seeing of essences':

> According to what has been stated, deductive theorizings are excluded from phenomenology. *Mediate inferences* are not exactly denied to it; but, since all

its cognitions ought to be descriptive, purely befitting the immanental sphere, inferences, non-intuitive moves of procedure of any kind, only have the methodic function of leading us to the matters in question upon which subsequent direct seeing of essences must make given.

(1913, p. 169)

The twin commitments of Husserlian phenomenology to be purely descriptive of experience and to be a science of essence seem to commit it to the notion of eidetic intuition, or the seeing of essences. Of course, Husserl does not suppose that I may see an essence in exactly the same sense that I may see a tree. He has a broader notion of 'seeing' in mind. Spelling out exactly what this notion amounts to, however, is a significant task.

Things become particularly problematic if we focus more tightly on the relation between eidetic intuition and the aim of Husserlian phenomenology to be purely descriptive. An issue that arises here is the worry that Husserlian phenomenology is committed to the so-called 'myth of the given' (Bell, 1990, pp. 196–7). Sellars (1956) famously and influentially argued that, when it comes to perceptual knowledge, the traditional distinction between brute, non-conceptual sensory input received by the mind (the given), on the one hand, and that which the mind contributes (interpretation, categorisation, etc.), on the other, cannot be maintained. It is a myth to suppose that something merely given in experience could ever justify one in interpreting it or categorising it in any one particular way. Rather, if anything is to justify a perceptual judgement, it must itself be within 'the space of reasons'. Since it is non-conceptual the *given* is, almost by definition, incapable of providing such reasons, so there is no one best way to describe it. As McDowell puts the point:

> we cannot really understand the relations in virtue of which a judgement is warranted except as relations within the space of reasons: relations such as implication and probabilification, which hold between potential exercises of conceptual capacities. The attempt to extend the scope of justificatory relations outside the conceptual sphere cannot do what it is supposed to do.
>
> (1994, p. 7)

The idea here – a version of which is expressed in the latter half of Kant's famous dictum that '[t]houghts without content are empty, intuitions without concepts are blind' (Kant, 1781/1787, A51/B75) – is that knowledge must be based on *reasons*. But for something to be a reason to make a particular judgement with a certain content, it must justify the application of that content. But brute, non-conceptual sensory input cannot do this. It is, to use Kant's term, blind. If it is to justify perceptual judgement, perceptual experience must already be conceptually structured. This claim should sound familiar as it is a variation on the already encountered view, articulated by Heidegger, that '[i]nterpretation is never a presuppositionless grasping of something previously given' (1927a, p. 146). Heidegger, it seems, does not recognise the given.

On the face of it, however, Husserl's demand for pure description, devoid of any bias or interference that may come, for example, from the tacitly held beliefs of the natural attitude, is indeed committed to just such a notion of the given. After all, Husserl's aim is to describe phenomena purely *as they are given*, in a way untainted by the prejudices of the natural attitude and uninfluenced by theories of the mind that rely on the postulation of Objects. If the given really is a myth, then, on the face of it at least, Husserlian phenomenology seems to be up to its neck in it (cf. Soffer, 2003).

We shall return to the alleged myth of the given when discussing perception in Chapter 3. The issue is worth mentioning here, however, since there is some reason to suppose that if the given is problematic, then the problem affects not just theories of sense experience but also views, such as Husserl's theory of eidetic intuition, that countenance non-sensory intuitive experience. Husserl is committed to the view that we may *see* essences and, on that basis, describe them, thereby achieving knowledge of the nature of phenomena. But, if Sellars is right that no brute input could ever by itself justify a particular judgement about it, then eidetic intuition would be mysterious indeed. However Husserl's notion of a non-sensory, eidetic intuition is to be spelled out, its defenders must meet this challenge, perhaps in the Heideggerian fashion mentioned above.

3.2 Heideggerian phenomenology

Whilst Husserl's methodological strictures set the agenda for much work within the Phenomenological tradition, they can hardly be said to have been uncritically endorsed by subsequent phenomenologists. Far from it, and among the various critics of the letter of Husserlian phenomenological method, it is Heidegger's work that stands out as of central importance. The relationship between Heidegger's early works and Husserlian phenomenology is both complex and deeply contested, with some commentators (e.g. Crowell, 2004) arguing that many central Husserlian claims are taken over by Heidegger, and others (e.g. Carman, 2003, Ch. 2) arguing, to the contrary, that Heidegger's philosophy represents a wholesale rejection of Husserlian phenomenology. This complexity is mirrored in the personal relationship between the two men, perhaps exemplified by the fact that whilst dedicating *Being and Time* (1927a) to Husserl 'in friendship and admiration', in 1926 Heidegger was writing to Karl Jaspers that '[i]f the treatise is written against anyone, it's against Husserl' (Letter to Jaspers, 26th December 1926. Quoted in Husserl, 1997, p. 28).

Since, in this chapter, our concern is with phenomenological methodology, we may ask how the early Heidegger conceived of phenomenology and, in particular, how he viewed the various aspects of Husserl's position that have been articulated above. At a certain point in the Introduction to *Being and Time*, Heidegger strikes what may seem a familiar Husserlian note:

> The term 'phenomenology' expresses a maxim that can be formulated: 'To the things themselves!' It is opposed to all free-floating constructions and accidental findings; it is also opposed to taking over concepts only seemingly demonstrated; and likewise to pseudo-questions which often are spread abroad as 'problems' for generations.
>
> (1927a, p. 26)

But while this indeed sounds Husserlian – and it would be a mistake, I think, to suppose that Heidegger and Husserl share nothing on the methods appropriate to phenomenology – it hides within it the seed of a fundamental disagreement. For the refusal to take over concepts 'only seemingly demonstrated' is a weapon that Heidegger wields against Husserl himself.

There are a number of ways in which Heidegger arguably departs from Husserlian phenomenology (his understanding of which is summarised in Heidegger, 1925, §10). Key among these is something that I mentioned above, his rejection of Husserl's demand that phenomenology be presuppositionless, employing purely descriptive methods. In seeing why Heidegger rejects this picture, replacing it with one that has become known as 'hermeneutic phenomenology', we can begin to appreciate Heidegger's enormously influential account of phenomenological method.

As we have seen, Husserl claims that the phenomenological reduction allows us to focus on phenomena, thereby enabling us to describe them in a way that is neutral with respect to the reality of the worldly objects around us. Heidegger complains, however, that such a procedure is far from neutral. On the contrary, it presupposes a distinction between the immanent and the transcendent that is not itself phenomenologically grounded, but taken over from the philosophical tradition, notably Descartes (Derrida, 1967, also complains that Husserl uncritically adopts traditional metaphysical concepts). In Heidegger's words:

> The elaboration of pure consciousness as the thematic field of phenomenology is *not derived phenomenologically by going back to the matters themselves* but by going back to a traditional idea of philosophy.
>
> (1925, p. 107)

As such, it can be objected that the distinction between the immanent and transcendent – a distinction upon which the formulation of Husserlian phenomenology relies – cannot be investigated from within the confines of Husserlian phenomenology. Husserl's view, according to Heidegger, simply assumes that, for example, the field of phenomenology is to be identified with that which is absolutely given. But not only does this and other assumptions mean that the Husserlian enterprise is far from presuppositionless, it is also symptomatic of a neglect of what Heidegger considers to be the most fundamental task of phenomenology: the question of the meaning of being.

The guiding question of Heidegger's work 'is to work out the question of the meaning of "*being*"' (Heidegger, 1927a, p. xxix); the question of what it is for something, of a given sort, to be rather than not be. On Heidegger's reading, Husserl's view as outlined above bypasses exactly this crucial issue. In setting up the domain and methods of phenomenology in the way that he does, Husserl does not so much as investigate the being of consciousness. Rather, he makes assumptions about what it is for consciousness to be, guided not by consciousness itself, but by 'the following concern: *how can consciousness become the possible object of an absolute science?*' (Heidegger, 1925, p. 107). Husserlian phenomenology, thinks Heidegger, is blind to the question of the meaning of the being of consciousness itself.

But what exactly is the question of the meaning of being, with which Heidegger is concerned? This is a vexed question within the interpretation of Heidegger's philosophy, and one to which there is no easy answer. Commentators have differed widely on what Heidegger means by 'being'. On Dreyfus's influential reading, for example, being is the 'intelligibility' of entities and Heidegger's goal is to 'make sense of our ability to make sense of things' as the things they are (Dreyfus, 1991, p. 10; also see Carman, 2003, Ch. 1). Blattner, on the other hand, complains that identifying the being of an entity with its intelligibility, or our 'ability to make sense' of it, turns Heidegger into a dogmatic idealist. Instead he interprets Heidegger as claiming that being is precisely that which we have an ability to make sense of. Being is not the intelligibility of things but that which is made sense of when they are (Blattner, 1999a, Introduction; also see McDaniel, forthcoming).

Whilst this interpretational issue is crucial for an understanding of the substance of Heidegger's philosophical project, it can be set aside for the purpose of appreciating Heidegger's objections to Husserlian phenomenology. For whether or not Heidegger identifies being with intelligibility, it is clear that he does think that we *do have* an ability to make sense of (the being of) things; we have what he calls a 'pre-ontological understanding of being'. He writes:

> We are able to grasp beings as such, as beings, only if we understand something like *being*. If we did not understand, even though at first roughly and without conceptual comprehension, what actuality signifies, then the actual would remain hidden from us. If we did not understand what reality means, then the real would remain inaccessible.
>
> (1927b, p. 10; also see §20 of the same work)

Heidegger's claim here is that for something to be 'accessible' to us as such, we must already possess some understanding of it. Thus, to ask after the meaning of being is possible only because we already have some level of understanding of being: 'I can comport toward beings only if those beings can themselves be encountered in the brightness of the understanding of being' (1927b, p. 281). This is a point familiar from Plato's *Meno* (2010), in which the worry is expressed that inquiry is either pointless or impossible since we either already know what we are looking for (so pointless) or we do not (so impossible). It is also something that we have already come across in §2.4, in the form of an objection to Husserl's account of variation in imagination.

In Heidegger's hands, the necessity of such a condition on the possibility of phenomenological, or any other, enquiry, shows that phenomenology must be interpretive rather than purely descriptive. Phenomenology involves the interpretation of phenomena on the basis of such a prior understanding of their being. Understanding, on Heidegger's picture, is a skill or form of 'know-how' (1927b, p. 276). Thus, Heideggerian understanding is something like what we have in mind when we say that one understands how to play chess, or ride a bicycle (cf. Ryle, 1949, Ch. 2). This prior understanding is a necessary condition of something's being there for us:

> [being] is understood as yet pre-conceptually, without a logos; we therefore call it *the pre-ontological understanding of being* [...] experience necessarily presupposes a pre-ontological understanding of being as an essential condition.
>
> (Heidegger, 1927b, p. 281)

What might such a pre-ontological understanding of being look like? Consider once more the tree that I see beyond my window. Heidegger's claim is that for the tree to show up for me at all, I must have some understanding of what it is for there to be a tree there. This, in turn, is a matter of my being able to engage with the tree in an understanding fashion – I can lean on it, climb it, lie under it in summer, rake its leaves in autumn, and so on. This collection of capacities, or know-how, constitutes my implicit understanding of what it is for there to be a tree in my garden and it is a necessary condition of the tree's forming a part of my *world*.

According to Heidegger, phenomenology – in fact, all forms of systematic enquiry – involves making this implicit understanding of the being of entities explicit. As he says, 'the methodological meaning of phenomenological description is *interpretation*' (1927a, p. 35). This making explicit of what is already implicitly grasped in understanding is what Heidegger calls interpretation, remarking that

> Interpretation does not, so to speak, throw a 'significance' over what is nakedly objectively present and does not stick a value on it, but what is encountered in the world is always already in a relevance which is disclosed in the understanding of world, a relevance which is made explicit by interpretation.
>
> (1927a, p. 145)

Again, in a comment surely directed to Husserl's own understanding of phenomenology as a pre-suppositionless science,

> Interpretation is never a presuppositionless grasping of something previously given. When a specific instance of interpretation (in the sense of a precise textual interpretation) appeals to what 'is there', then that which initially 'is there' is nothing other than the self-evident, undiscussed prejudice of the interpreter which necessarily lies in every interpretive approach as that which is already 'posited' with interpretation in general, namely that which is pre-given in fore-having, fore-sight, and fore-conception.
>
> (1927a, p. 146)

How does this Heideggerian account of phenomenology as the explication of what is already implicitly understood relate to the various elements of Husserl's method? In particular, what can be said about both the phenomenological and eidetic reductions?

With respect to the phenomenological reduction, we have already seen that Heidegger queries Husserl's implicitly assumed conception of the transcendent that plays a role here. But there is another respect in which the account out-lined above might be thought to call the phenomenological reduction into question. For, on Husserl's account, the reduction involves the bracketing of Objects, suspending our belief in their existence. But if the pre-ontological understanding upon which interpretation is based is a *practical* rather than a cognitive form of understanding – a kind of know-how – then it is not clear that there *is* such a belief operative in the natural attitude. This is a reading suggested by the following remarks from Dreyfus:

> we cannot get clear about the 'beliefs' about being that we seem to be taking for granted. There are no beliefs to get clear about; there are only skills and practices [...] we can only give an interpretation of the interpretation already in the practices [...] an explication of our understanding of being can never be complete because we dwell in it.
>
> (1991, p. 22)

In the face of this, however, the Husserlian might point out that even if the natural attitude does not involve belief, so there is no belief for the phenom-enologist to suspend, the purpose of the phenomenological reduction is in any case to free us from the ontological commitments of the natural attitude. The fact, if it is one, that this is not to be done by suspending a belief in Objects does not really cut to the heart of the reduction.

This does not let the Husserlian off the hook, however, since Heidegger's point is that there is a more profound sense of ontological commitment in the natural attitude in its conception of the being of consciousness and worldly entities than cannot be neutralised by any bracketing. On this reading, the problem with the phenomenological reduction is not that there is no belief to suspend but that the practical nature of understanding entails that the suspension of belief is insufficient to insulate the phenomenologist from ontological commitment.

What of the eidetic reduction? Once more, we have already seen a crucial respect in which the Heideggerian challenge to Husserlian phenomenology calls the reduction into question: the claim that the associated method of variation in imagination is not neutral but rests on prior implicit conceptions of the nature of the domain under phenomenological investigation. It should be emphasised, though, that whilst Heidegger has critical things to say about the eidetic reduction, he is in harmony with Husserl on its central outcome. He agrees, that is, that phenomenology is an *a priori* discipline that seeks the essences of phenomena. In phenomenology, '[n]ot arbitrary and accidental structures but essential ones are to be demonstrated' (1927a, p. 16). Thus – and this relates also to Husserl's view of the significance of the phenomenological reduction – phenomenology cannot rely on the deliverances of the empirical sciences but must proceed after its own fashion. In this, Husserl and Heidegger are in firm agreement.

There is agreement, then, between Husserl and Heidegger regarding the priority of phenomenological investigations over those of the positive sciences (cf. Heidegger, 1927a, §3). But there are deep divisions in their respective understandings of phenomenology. It is fair to say that these divisions were fundamental in the development of Phenomenology throughout the twentieth century. Whilst Husserl is universally acknowledged as the founder of phenomenology, it can certainly be argued that Heidegger's conception of what phenomenology must be has been at least as influential.

4 Conclusion

Sartre's excitement at the prospect of engaging in a broadly Husserlian phenomenology might be understood as motivated, at least in part, by a desire to leave behind seemingly intractable disputes in metaphysics and epistemology, embarking on a radically modern approach to philosophy – the rigorous study of the essences of phenomena. Whilst a broad conception of phenomenology sees its location within a range of wider philosophical concerns, narrowly construed it is a science of things *as they appear*. On Husserl's understanding, this involves the adoption of both phenomenological and eidetic reductions. There are, however, a number of potential objections to Husserl's prescriptions for phenomenological method, the most influential of which is Heidegger's critique. We are left, then, with an unresolved dispute between rival ways of understanding the nature of phenomenology. We should, however, have sufficient understanding to enable us to follow the advice of both Husserl and Heidegger when they tell us to go back to the things themselves!

2   The objects of experience

'Consciousness of something' is therefore something obviously understandable of itself and, at the same time, highly enigmatic

Husserl, *Ideas I*

It is through experience that we engage with the world around us. At any given moment each of us enjoys sensory experiences in the various modalities (visual, auditory, tactile, etc.); we accept many things as true of the Objects we experience; we feel joyful, sad, angry, or afraid of them, and so on. In each case, so it seems, we are in contact with things in the surrounding world; in each case an Object is *there for us.* Arguably the primary goal of anyone working within the Phenomenological tradition is to give an account of this fundamental feature of experience, its making it such that things are *there for us.* It is this feature of experience that is addressed by accounts of intentionality and it is to this notion that we turn in the present chapter.

1 Intentionality in the Phenomenological tradition

It is only a slight exaggeration to say that the history of the Phenomenological tradition is a history of the various interpretations and perceived significance of the concept of intentionality. Brought to prominence by Brentano, elaborated by Husserl, employed and modified in various ways by Heidegger, Sartre, and Merleau-Ponty, intentionality is fundamental to the phenomenological enterprise. Discussions of perception, imagination, self-consciousness, and emotion are shot through with assumptions, claims, and counter-claims about the nature and scope of intentionality. The focus on intentionality is one of the Phenomenological tradition's most enduring legacies, finding its way into almost every aspect of philosophical work on the mind.

Brentano is typically credited as the origin of the contemporary debate. His concern, in *Psychology from an Empirical Standpoint*, was to demarcate the subject matter of psychology or, as he also puts it, to distinguish 'mental phenomena' from 'physical phenomena', where *phenomena* are just things that appear to, or are perceived by, us. He argued that the feature that best characterises mental phenomena is their *intentional inexistence*. In a now famous passage, he wrote:

> Every mental phenomenon is characterised by what the Scholastics of the Middle Ages called the intentional (or mental) inexistence of an object, and what we might call, though not wholly unambiguously, reference to a content, direction towards an object (which is not to be understood here as meaning a thing) or immanent objectivity. Every mental phenomenon includes something as object within itself, although they do not all do so in the same way. In presentation, something is presented, in judgement something is affirmed or denied, in love loved, in hate hated, in desire desired and so on.
>
> (1874, p. 88)

Less concerned with the question of whether intentionality is the defining characteristic of the subject matter of psychology (the 'mark of the mental'), Husserl nevertheless took over the notion of intentionality, making it foundational for his phenomenology generally, and for his various analyses of specific phenomena. Returning time and again to the structure of intentionality, his account sets the agenda for subsequent developments within the Phenomenological tradition. These subsequent developments are complex but one persistent theme, originating with Heidegger's account of being-in-the-world and continuing in the work of Merleau-Ponty, is the view that the most basic form of intentionality (or directedness) is practical, what Merleau-Ponty calls 'motor intentionality' (see Kelly, 2000). This is often presented as an objection to the Husserlian account, the thought being that Husserl's account of intentionality is inadequate to the phenomenology of what Dreyfus has influentially called 'absorbed coping' (1993; also see Rouse, 2000). We shall return to this issue in §4. To begin with, however, the focus will be on the accounts presented by Brentano and Husserl.

2 The mark of the mental

As Brentano says, intentionality is 'reference to a content' or 'direction towards an object'. (He rather confusingly uses 'content' and 'object' interchangeably. We will return to this point later.) Consciousness, or experience, is experience *of something*. This simple thought has been the bedrock of much theorising about the mind, in both philosophy and psychology, since Brentano was writing towards the end of the nineteenth century. In the next section we will look in more detail at the structure of intentionality itself. First, however, it is useful to consider Brentano's Thesis that all and only mental phenomena exhibit intentionality.

Brentano wanted a general way of distinguishing between physical and mental phenomena. His examples of mental phenomena include hearing, seeing, feeling, imagining, thinking, judging, recollecting, expecting, inferring, doubting, willing, intending, desiring, admiring, joy, and hatred. His examples of physical phenomena are the objects of the senses and include: colours, figures (shapes), chords (sounds), heat, cold, and odours. Brentano considers a number of different accounts of what unites all and only mental phenomena, rejecting some and accepting others. For example, he thinks that it is true that all and only mental phenomena 'are either presentations or they are based upon presentations' (1874, p. 85). He does not, however, consider this account to be entirely satisfactory since it characterises the category of mental pheno-mena disjunctively. Better, he thinks, is the view that all and only mental phenomena exhibit intentionality.

Mental phenomena appear to us, according to Brentano, through 'inner consciousness', but it is *through them* that physical phenomena appear to us. So, whilst both mental and physical phenomena appear to us, it is only the former that themselves have objects. As a handy shorthand, I will use 'inten-tional object' for 'object of an intentional experience'. Thus, the claim is that all and only mental phenomena have intentional objects. This view is, in fact, closely related to the account in terms of *presentation* since, according to Brentano's usage, to 'be presented' is 'to appear', and the object of a pre-sentation (or an experience founded on a presentation) is simply that which appears. This also brings into view the relation between Brentano's account of mentality and Husserl's account of phenomenology, as introduced in Chapter 1. According to Husserl, the aim of the science of experience is to describe both experience and *things as they appear in experience*. Thus, given Brentano's Thesis, another way to put this would be to say that phenomenology seeks to describe both experience and its objects just as they are given. Intentionality, it seems, is implicated in the very description of what phenomenology seeks to achieve.

Returning to Brentano's view, we want to know whether intentionality is indeed both necessary and sufficient for mentality. A natural objection to Brentano's Thesis is that words, sentences, propositions, novels, paintings, photos, etc. often refer to, or are directed towards, objects. Yet none of these things are mental. If this is right, then intentionality cannot be *sufficient* for mentality.

One response to this objection is to distinguish between natural and con-ventional intentionality (or intrinsic and derived intentionality, cf. Searle, 1983). Words and pictures, for example, are only *conventionally* directed towards objects. In order for there to be some object to which they refer, there must be someone who *uses* the sentence to so refer. The conventional intentionality of, for example, words and pictures, depends on the existence of other intentional phenomena, the intentionality of which is not conventional in this way. These are the mental phenomena. The directedness of, for example, a thought towards its object is *natural*. In order for my thought to refer to its object, I need not *use* it to do so. Indeed, it is difficult to know what such a using would amount to. Thoughts, and other mental phenomena, need no help in referring to their objects, they do so all by themselves. If the details of such a distinction can be spelled out in a satisfactory way, Brentano's Thesis might be restated as claiming that a phenomenon is mental if and only if it exhibits natural, or undervied, intentionality.

A potentially more serious objection to Brentano's Thesis is that intentionality is not *necessary* for mentality since there are some non-intentional experiences. The most common examples offered are bodily sensations such as pains or orgasms, and certain types of undirected emotions or moods, such as nervousness, anxiety, or depression.

It is plausible enough that beliefs, perceptual experiences, and most emotions are directed towards objects: I believe that the *Eiffel Tower* is in *Paris*, I see *the tree*, I am afraid of *the spider*. But what is the intentional object towards which a headache is directed? Why should we suppose that it refers to an object at all? Is it not more plausible to suppose that pain is entirely subjective, a pure feel, lacking in directedness? Similar questions arise for cases of undirected nervousness and other moods. Searle, for example, maintains that 'there are forms of nervousness, elation, and undirected anxiety that are not Intentional' (1983, p. 1). If correct, this view would upset Brentano's account of the defining feature of mental phenomena.

Brentano's own answer to this concern, in the case of bodily sensations at least, is that the intentional object of sensation is a sensory quality instantiated by a spatially located body part. As he points out, in sensation

> we always have a presentation of a definite spatial location which we usually characterize in relation to some visible and touchable part of our body. We say that our foot or our hand hurts, that this or that part of the body is in pain. [...] there is in us not only the idea of a definite spatial location but also that of a particular sensory quality analogous to colour, sound and other so-called sensory qualities, which is a physical phenomenon and which must clearly be distinguished from the accompanying feeling [...] in cases where a feeling of pain or pleasure is aroused in us by a cut, a burn or a tickle, we must distinguish in the same way between a physical phenomenon, which appears as the object of external perception, and the mental phenomenon of feeling, which accompanies its appearance, even though in this case the superficial observer is rather inclined to confuse them.
>
> (1874, pp. 82–3)

To this Brentano adds that we are led to wrongly suppose that pain is purely subjective by the fact that we tend to use the word 'pain' to refer to both the state of bodily awareness and the quality of the body of which we are aware. Strictly speaking, however, pain is the awareness of the body as possessing the sensory quality in question. It is thus wrong, albeit natural, to suppose that pains are, literally, located in parts of the body.

There is much to be said about this account of pain but, for present purposes, it will suffice to note that some story along these lines is required of anyone wishing to claim that intentionality is a necessary condition of the mental since, if anything is to count as a mental phenomenon, surely it is pain. We will have more to say about pain and other bodily sensations in Chapters 7 and 8.

What of undirected moods as putative counterexamples to Brentano's Thesis? Whilst some emotions, such as fear or remorse, seem to exhibit intentionality, many have supposed that certain moods do not. If one is in a good mood, or is depressed, one need not be in a good mood, or be depressed, about anything in particular. That is not to say that we never think of these experiences as directed. We might, for example, be depressed about the state of political discourse, or about our career prospects. But there seem to be easily recognisable cases in which one is just depressed and that's that. Unlike paradigm cases of emotions that one is in for some reason – I'm happy because I won a prize – one is not in undirected moods for any *reason* (which is not to say that they are not caused by something or other). These seem to be experiences that simply wash over one, and are directed at nobody and nothing in particular.

We will return to emotions and moods in Chapter 10. Here I will simply mention the sort of answer that proponents of Brentano's Thesis might offer. This is that depression, anxiety, and other moods, are not undirected at all. They only seem so because they are in fact directed towards very general objects. The intentional object of my depression may be the world as a whole, or perhaps my position within it. As Heidegger puts it, employing a characteristic phrase, '*That in the face of which one has anxiety is Being-in-the-world as such*' (1927a, p. 180; cf. Crane, 1998).

3 The structure of intentionality

Both Brentano and Husserl employ the term 'phenomenon' to mean, roughly, that which appears. Since phenomenology is the science of phenomena, we might expect it to be concerned with, in Brentano's terms, both mental and physical phenomena. And so it is. Phenomenology is concerned to describe both experience and the objects of experience *as they are given*. In many cases, these will be phenomena that are not themselves intentional. The Husserlian phenomenologist, then, is concerned with Brentano's Thesis not because it serves to delimit the subject matter of phenomenology – it does not; the phenomenological reduction supposedly does that – but because if true it tells us something crucial about the nature of experience. Thus far, however, we have provided little detail concerning the nature of intentionality itself, simply getting by with the rough idea that intentionality involves directedness to an object. Both Brentano and, in particular, Husserl have a great deal more to say on this score.

Before getting into some of the details, it should be recalled that, this being a phenomenological investigation, the task is purely descriptive. We are not concerned, for example, with the question of whether intentionality can be accounted for in purely physical, or otherwise naturalistically acceptable, terms. That project, as worthwhile as it may be, cannot be our initial concern. In order to answer that question in a meaningful way, we must first have a clear grasp of what intentionality is. Our understanding of intentionality, based as it is on our first-hand experience, must be what guides any subsequent investigation as to how it may or may not fit into the natural world. Our goal, then, is to describe the structural features of intentionality. We will begin with the nature of intentional objects, a question over which, it would seem, Brentano and Husserl sharply disagree. In fact, Brentano's view of this matter is difficult to interpret.

3.1 Intentional objects

According to Brentano, mental phenomena are those that refer to an object. To use one of his own examples, *hearing a sound* is a mental phenomenon, the *chord* heard is its intentional object. Whilst the hearing of the sound is mental, the sound/chord is a physical phenomenon, since it contains no reference to any further object. But what sort of thing is an intentional object? One could certainly be forgiven for supposing it to be a physical thing. On this understanding, the objects of intentional experiences would be Objects in Husserl's sense, entities the reality of which transcend our consciousness of them. This interpretation, which we can call the Object view, would make good sense of Brentano's stated project of demarcating the subject matter of psychology, as distinct from other scientific enterprises.

It is, however, difficult to read Brentano as consistently maintaining this view. In his celebrated introduction of the concept of intentionality, quoted above, he includes the somewhat surprising remark that the object to which an intentional experience is directed, 'is not to be understood here as meaning a thing' (1874, p. 88). A clue as to the explanation of this remark can be gained by consulting Brentano's lecture series *Descriptive Psychology*, in which he writes that '[b]y phenomena, however, [I understand] that which is perceived by us, in fact what is perceived by us in the strict sense of the word', and that '[t]his, for example, is not the case for the external world' (1890–1, p. 137). This 'strict sense' is associated, by Brentano, with indubitability (1874, p. 91). Thus, the thought seems to be that since the existence of the world of Objects is open to doubt, Objects cannot be phenomena.

What, then, are phenomena? One way to make sense of Brentano's position is to claim that his doctrine of *intentional inexistence* holds that an experience's intentional object is literally a part of the experience itself, existing *in* it (Crane, 2006a). After all, Brentano does tell us that each 'mental phenomenon includes something as object within itself' (1874, p. 88). Not only does this interpretation make sense of the above quotations, it is also motivated by the view that Brentano was working with essentially the same indirect view of our relation to the external, Objective, world as is often attributed to the British Empiricists, including Locke and Hume (Bartok, 2005). It is often claimed that, in his *Essay Concerning Human Understanding*, Locke held the view, crudely expressed, that our access to the external world is not direct but mediated by what he called *ideas* which are not Objects but rather subjective entities that are, in some sense, in the mind. That is, in seeing a tree I am not in immediate contact with the tree itself, but rather with an idea of the tree. A common concern about this view is that this 'veil of ideas' would have the effect of cutting us off from the surrounding world, ushering in scepticism about the 'external world'. We shall encounter this view once more in the following chapter, in the discussion of perceptual experience. For now the point is simply that, if Brentano accepted such a view, one would certainly expect him to suppose that phenomena are distinct from Objects.

One benefit of this account of the nature of intentional objects is that it has no difficulty in dealing with experiences that seemingly refer to things that do not in reality exist. One may be aware of an idea of a tree even if there is no corresponding Object, i.e. no real tree. This is not so with the view that intentional objects are Objects. If I think *Obama is President*, the Object view will maintain that the object of my thought is Barack Obama, the man himself. If, on the other hand, I think *Atlantis does not exist*, the Object view cannot say that the object of my thinking is Atlantis, the place itself, for there is no Atlantis. The Object view, then, needs to explain how it is that I can think about things that do not exist (note here that, in order to raise this issue, we have stepped outside of the phenomenological reduction).

On the face of it, a proponent of the Object view has a number of options in dealing with such cases. First, they may deny that my thinking is about an object at all. Of course, given Brentano's Thesis, this would disqualify my thinking from the category of mental phenomena, surely an unwelcome result. Second, they could try to find some suitable Object that can be the intentional object of my thinking. In the case at hand, since 'Atlantis' names a fictional nation, it might be held that it refers to a special type of abstract entity, a *fictional object*. Third, they might make a distinction between existent objects (that have being) and non-existent objects (that have non-being), and claim that Atlantis is a non-existent object. On such a view, a version of which was held by Meinong (1904), intentional objects need not exist. Of course, such a view would need to say a great deal more about the distinction between having existence and having non-existence before it could be made plausible. Brentano, for one, confesses that he is 'unable to make any sense of this distinction' (1874, p. 274). Many philosophers will consider this an option of last resort.

The view according to which phenomena are not Objects, but rather parts of the intentional experiences themselves, can easily avoid this problem. For whenever an intentional experience exists, we can guarantee that its object will. This certainly speaks in its favour. There are, nevertheless, puzzles for it.

For one thing there is the concern that it serves to sever us from the Objective world. Indeed it does so in a quite radical way since, given that no intentional experience would have an Object as its object, we would not even be in a position to *entertain the thought* of an Object. Thus, despite appearances, none of the considerations in this discussion would be about Objects at all, not even this one!

Second, this view has a more difficult job of explaining Brentano's distinction between mental and physical phenomena since he clearly thinks that some things that he calls physical phenomena are intentional objects. But if such things are themselves parts of intentional experiences it is difficult to avoid the thought that they are part of the subject matter of psychology, thus conflicting with Brentano's own categorisation.

Finally, Brentano tells us that the objects of some intentional experiences do not exist, using this fact as reason to deny that intentionality is a relation, since relations require all of their relata to exist. As he puts it:

> If someone thinks of something, the one who is thinking must certainly exist, but the object of his thinking need not exist at all [...]. The terminus of the so-called relation does not need to exist in reality at all. For this reason one could doubt whether we really are dealing with something relational here.
>
> (1874, p. 272)

On the face of it, this contradicts the supposed benefit of the view that it guarantees an actually existing object for every intentional experience. It also re-raises the question that dogs the Object view, of how an experience could refer to an object that does not exist. By the same token, however, Brentano's suggestion that intentionality is not a relation may hold the key to answering this puzzle, and it suggests that his view of intentional objects is in some respects pointing towards the account offered by Husserl.

In *Logical Investigations* Husserl interprets Brentano as claiming that the object of an intentional experience is literally contained within it. Whether or not this is the right way to interpret Brentano, it is a view that Husserl adamantly rejects. The object of an intentional experience, thinks Husserl, is no part of it. He offers at least two lines of thought against the view. The first is that it rests on a conflation of the objects and the 'contents' of experience. I return to this below in the discussion of sensation. The second is an argument against what he calls the 'image theory'. This is a version of the view, typically attributed to Locke and Hume and, as we have seen, perhaps held by Brentano, according to which our awareness of Objects in the surrounding world is indirect, mediated by an awareness of ideas or 'images' of them. His argument (in Husserl, 1900–1, Vol. 2, Appendix to §11 and §20) runs loosely as follows: A is only an image of B if it is interpreted as such. But for A to be interpreted as an image of B, both A and B must be given to consciousness. It cannot, therefore, be that all experiences have only images of Objects as their intentional objects. Without the capacity to represent Objects directly, nothing could so much as be an image of an Object since nothing could be interpreted as an image of it.

We have, however, in the distinction between natural and conventional intentionality, already seen some reason to doubt that this argument hits its target. Images exhibit conventional intentionality and so, as Husserl points out, in order to be directed to some object they must be so interpreted or used. But the view that *conventional* intentionality is the mark of the mental is not plausible. Rather, the view must be that experiences exhibit *natural* intentionality. As such, Husserl's argument against the image theory does not give us a reason to reject Brentano's position. Another way of putting the point is to say that the argument rests on a special feature of images – that they are conventional – that need have no place in the view that intentional objects are parts of intentional experiences.

In any case, Husserl does not rest much on the critique of the image theory. In fact, he seemingly takes the Object view as obvious, describing the following remarks as 'truisms':

> It need only be said to be acknowledged *that the intentional object of a presentation is the same as its actual object, and on occasion as its external object, and that it is absurd to distinguish between them.* The transcendent object [Object] would not be the object of *this* presentation, if it was not *its* intentional object [...]. If I represent God to myself, or an angel, or an intelligible thing-in-itself, or a physical thing or a round square etc., I mean the transcendent object [Object] named in each case, in other words my intentional object: it makes no difference whether this object exists or is imaginary or absurd. 'The object is merely intentional' does not, of course, mean that it exists, but only in an intention, of which it is a real part, or that some shadow of it exists. It means rather that the intention, the reference to an object so qualified, exists, but not that the object does.
>
> (1900–1, Vol. 2, p. 127)

This evidently is the Object view. The intentional object of an experience is a transcendent entity, not a part of the experience itself. In cases such as that of thoughts about Atlantis, or a round square, we simply say that whilst the experience exists, its object does not.

How, then, does Husserl answer the above puzzle about experiences whose objects do not exist? How can an experience have Atlantis as its object if Atlantis does not exist? Husserl is clear that intentional objects are Objects. But, if intentional objects are what intentional experiences are directed towards, and some intentional experiences are about things that do not exist, then some intentional objects do not exist. If so, they surely cannot, in every case, be ordinary Objects.

The answer to this puzzle perhaps lies, as suggested above, in the recognition that intentionality is not a relation to an Object, and an associated view about what determines the intentional object of any given experience. Despite the way that Husserl and others sometimes speak, there is, as Brentano recognised, reason to suppose that intentionality is not a relation to an Object. Rather, that a given experience refers to some object is determined by features internal to it (and perhaps to the abstract *matters* or *senses* introduced in the next section). This point might be made clearer by placing intentional directedness alongside an intuitive contrast between pointing and speaking.

If I point my finger, I may succeed in picking out some object. What object that is will depend on the layout of the world around me. I may think that I am pointing at a tree but in fact be pointing at a picture of a tree, or at nothing at all. Thus, that a given instance of pointing picks out some object is determined not just by features internal to the pointing (the arrangement of my fingers) but also by other things (the arrangement of the Objects around me). Contrast this with an intuitive thought about speaking. If I utter 'Barack Obama', I pick out a person, Barack Obama. There is no obvious way in which my utterance will fail to pick out Obama and instead refer to someone else. That my utterance refers to Obama, it would seem, is determined by factors internal to it. If Obama did not exist – if 'Obama' were an empty name – my utterance would still have a sense, I would have uttered something whose function it is to refer to the actual Obama. So whilst for my pointing to have sense, there needs to be something towards which I point, this is not so for speaking. In saying that intentionality is not a relation to an Object, one is saying that in this respect it is closer to speaking than to pointing.

Now, what I have said above about speaking is both highly controversial and brushes over a number of very important distinctions that philosophers of language will wish to draw. It embodies a commitment to *semantic internalism*, the view sometimes put by saying that 'meaning is in the head'. The internalist claims that what my utterances mean, their 'content', is determined by facts internal to me and depends on nothing external (Segal, 2000). The externalist, by contrast, denies that meaning is in the head. Rather, according to externalism, for my utterances to mean what they do I must stand in appropriate relations to the relevant Objects (Putnam, 1975). Stated even in this crude way, we can see that, on the internalist picture, it does not matter whether the Object of one's utterances really exists, whereas on the externalist picture it does. Treating intentionality as a relation to an Object is, in effect, assuming a form of externalism. If, on the other hand, Husserl's account of intentionality embodies a commitment to internalism, this would explain why he is not overly concerned with the problem of non-existent objects.

That Husserl *would* adopt an internalist account of intentionality is to be expected given his general account of phenomenological method, discussed in the previous chapter. There I noted that Husserl's claims regarding the phenomenological reduction would be undermined by an acceptance of a view of experience known as naïve realism. Naïve realism is related to externalism, since it states that how things appear to me in experience depends not just on intrinsic features of experience, but on the relation of acquaintance that one bears to the Objects around one. Husserl's account of the way in which the phenomenologist should employ the phenomenological reduction presupposes the falsity of this picture and implicitly the acceptance of a form of internalism.

That Husserl *does* adopt an internalist account of intentionality is suggested by a number of comments that he makes. He tells us, for example, that '"reference to an object" belongs peculiarly and intrinsically to an act-experience' (1900–1, Vol. 2, p. 120) and that '[i]t is the act's matter that makes its object count as this object and no other [...]. Identical matters can never yield distinct objective references' (1900–1, Vol. 2, p. 122). We will look at Husserl's notion of 'matter' below. I note here, though, that his claim about the 'matter' of an experience yielding its reference recalls the Fregean slogan that 'sense determines reference' (Frege, 1892), and suggests that Husserl took it that which object an experience picks out is determined by features intrinsic to the experience itself. This, then, would be an internalist view.

If this is the right way to interpret Husserl, then one way of objecting to his account of intentionality is to argue against internalism. The most influential such argument is Putnam's (1975) *Twin Earth* thought experiment. We can imagine a world that is, in every way, like the actual Earth except for the fact that on this Twin Earth, there is no H_2O but rather an entirely distinct chemical kind, XYZ, that looks and tastes, etc. just the same as water and that inhabitants of Twin Earth call 'water'. It is plausible to suppose that when I, here on Earth, think *water is wet*, my thoughts are about water, i.e. H_2O. However, so the argument goes, when an inhabitant of Twin Earth with the same intrinsic properties as myself thinks *water is wet*, their thought is not about water but about twin-water, i.e. XYZ. If this is right, then it suggests that the intentional object of one's thought can depend on the Objects to which one is related. For one's thoughts to be about water, one must have had some interaction with water, i.e. H_2O. This is something that my twin on Twin Earth lacks.

Whilst Putnam's case for externalism has been very influential, it is not without its detractors. There are a number of ways to challenge externalism, of which I will briefly mention two. First, we can simply re-raise the problem of non-existent objects. If the capacity to think about water requires some form of interaction with water, then what are we to say of the case of Dry Earth (Boghossian, 1997), which contains no water but whose inhabitants are collectively under the illusion that there is? Should we say that their thoughts that *water is wet* lack an object at all? If one finds this implausible, one may be tempted to suppose that there must be something wrong with Putnam's argument. But where might the problem lie? One suggestion (Fodor, 1987) is that the argument fails to distinguish between two forms of content: whilst externalism is true of *wide content*, internalism is true of the *narrow content* of intentional states, narrow content being that which the thinkers on Earth, Twin Earth, and Dry Earth all have in common when they think *water is wet*.

This notion of 'content' looks ahead to the next section and Husserl's concept of the *matter* of an intentional state. But, putting aside the issue of the externalist challenge to Husserl's apparent internalism, we still want to know what intentional objects are, on Husserl's view. One answer is simply to say, with Husserl, that they are ordinary, transcendent Objects, whilst allowing that in some cases these do not exist. But such remarks have the air of paradox. Better, perhaps, to rely on the above discussion and offer a deflationary reading that replaces talk of intentional objects with talk about what intentional experiences purport to be about, or pick out. Since, given Husserl's internalist account of intentionality, an experience may purport to pick out something yet fail to actually pick anything out, intentional objects need not always be Objects and the phrase 'intentional object' (and the longer 'object of an intentional state') is, in this respect, misleading. Perhaps the best option is to drop the phrase 'intentional object' entirely. On this picture, then, the goal of phenomenology is to describe experiences and the things that they 'purport' to be experiences of, even if there are in fact no such things in reality.

3.2 Quality and matter

In *Logical Investigations* Husserl distinguishes between the intentional quality of a conscious act and its intentional matter. The distinction is between

> the general act-character, which stamps an act as merely presentative, judgemental, emotional, desiderative etc., and its 'content' which stamps it as presenting *this*, as judging *that* etc. etc. The two assertions '2 x 2 = 4' and 'Ibsen is the principal founder of modern dramatic realism', are both, *qua* assertions, of one kind; each is qualified as an assertion, and their common feature is their *judgement-quality*. The one, however, judges one content and the other another content. To distinguish such 'contents' from other notions of 'content' we shall speak here of the *matter* (material) of judgements.
>
> (1900–1, Vol. 2, p. 119)

The intentional quality of an experience is, then, its being a desire, visual perception, memory, and so on. This is plausibly an intrinsic feature of the experience itself. An experience's matter, on the other hand, is akin to the contemporary notion of the *content* of a judgement, desire, emotion, etc. This is, arguably, an abstract (or ideal) entity to which the experience is related. Each of these is, in turn, distinct from the intentional object. So, when I judge that the tree is a sycamore, the object of my act is the tree or, more precisely, it is the state of affairs consisting in the tree's being a sycamore; the quality is judgemental; and the matter is *the tree is a sycamore* (throughout I restrict 'judgement' to propositional judgements, i.e. judgings *that*).

There is more than a passing similarity between Husserl's notion of intentional matter and Frege's notion of sense. Indeed, Husserl refers to matter as the 'interpretative sense' (1900–1, Vol. 2, p. 122) of an experience. Frege famously distinguishes between the sense and reference of an expression (1892). Thus, whilst the reference of the expressions 'the morning star' and 'the evening star' is the same – i.e. the planet Venus – their sense differs. That these senses differ can be seen when we notice that a person can understand both 'the morning star' and 'the evening star' whilst quite coherently thinking something true of one that they consider false of the other, for example that it is visible in the evening. We can put this by saying that since the expressions have a different *cognitive significance*, they are associated with distinct senses.

The sense of an expression, on Frege's view, is the *mode of presentation* of its referent. Similarly, Husserl speaks of matters as 'ways of referring to objects' (1900–1, Vol. 2, p. 121), pointing out that 'the ideas *equilateral triangle* and *equiangular triangle* differ in content, though both are evidently directed to the same object: they present the same object, although "in a different fashion"' (1900–1, Vol. 2, p. 121). So, on Husserl's picture, matter is that feature of an intentional experience that

> *not merely fixes the object meant in a general way, but also the precise way in which it is meant* [...] not only determines *that* it grasps the object but also *as what* it grasps it, the properties, relations, categorical forms, that it itself attributes to it.
>
> (1900–1, Vol. 2, pp. 121–2)

If matter is akin to Fregean sense, the fact that Husserl thinks of every type of intentional experience – every quality – as possessing matter shows that it is a generalisation of that notion. We are to speak not only of the matter, or sense, of linguistic entities but also of perceptual experiences, memories, emotions, and so on. Thus, Husserl's account of intentionality as involving matter, quality, and object, offers us a unified and quite general account of the structure of intentional experience.

Frege might be criticised for leaving it unexplained how it is that we 'grasp' a sense (see, for example, Dummett, 1993, Ch. 10). That is, he offers no clear account of how it is that we concrete, minded beings existing in space and time are in touch with ideal meanings; abstract entities that do not exist within our minds or in space and time at all, but in what Frege called the 'third realm' (Frege, 1918). Indeed, Frege himself admitted of 'grasping' a sense that 'this process is perhaps the most mysterious of all' (1897, p. 246).

Husserl's account of intentionality in *Logical Investigations* is arguably clearer on this score. He claims that meanings are ideal species that are instantiated in the matters of intentional acts, just as universals are instantiated in worldly things (1900–1, Vol. 1, pp. 228–33). Thus, 'Meaning is related to varied acts of meaning [...] just as Redness *in specie* is to the slips of paper which lie here, and which all "have" the same redness' (1900–1, Vol. 1, p. 230). In this way, Husserl can account for the relation between we concrete, spatio-temporal people and the ideal meanings that we are in touch with – our intentional acts instantiate those meanings – in a way that avoids the metaphor of grasping (see McIntyre, 1987). Frege's mystery is, on this view, resolved. Of course, Husserl's account may be criticised (see, for example, Dummett, 1993, Ch. 6), but it is safe to say that it is a most penetrating and influential account of one of the deepest of philosophical questions: how is it that something can be *given* to someone at all?

3.3 Noesis and noema

In his *Ideas I* Husserl introduced a new terminology to describe the structure of intentionality, distinguishing between the *noesis* and the *noema*, claiming that phenomenology involves both noetic and noematic analysis (1913, Pt. 3, Ch. 3). The noesis is the conscious experience itself, including all those features that are 'really inherent in it which make it up' (1913, p. 213). This notion incorporates that of intentional quality, the latter being an intrinsic feature of experience. Thus, noetic analysis looks at the structure of conscious experiences and the different ways in which objects are consciously intended in them.

The noema, on the other hand, roughly corresponds to the earlier notion of intentional matter. Here, however, we encounter a significant dispute among interpreters of Husserl. The noema has been interpreted by some as the content of the intentional act – an abstract entity that, as with the earlier notion of matter, is akin to Fregean sense (Føllesdal, 1969) – and by others as the intentional object *as it is intended*, that is, considered from a particular perspective (Sokolowski, 2000). On the first understanding, noematic analysis looks at the meaning of intentional experiences, on the latter it considers objects *as they are given to consciousness*.

Exactly how to interpret Husserl's notions of the noema and noematic analysis is much debated (e.g. Woodruff Smith 2007, pp. 304–11). Each reading has a basis in Husserl's text. For example, Husserl speaks in a single breath of the noema of a perceptual experience as 'its perceptual sense, i.e. the *perceived as perceived*' (1913, p. 214). So, the noema is conceived as the sense, i.e. the matter, of an experience (indeed, in a footnote to the above quotation, Husserl directs the reader to the discussion of matter in *Logical Investigations*), but it is also conceived as, or as incorporating, the intentional object itself. But, as we saw in the previous section, Husserl distinguishes the matter of an intentional experience from its object. So, which is it to be? Is the noema equivalent to the matter of an experience, or to its object, or does it somehow incorporate both?

One way to resolve this puzzle might begin by pointing out that the noema is surely not the actual experienced Object since every experience possesses a noema, yet not every experience is directed towards an actual Object. The objectual view, then, must surely identify the noema with the *intentional* object. But, as I suggested in §3.1, the notion of an intentional object should be treated with caution. There I suggested that 'intentional object' is best understood as meaning 'what an intentional experience purports to be about'. But that an intentional experience purports to be about something is an aspect of its sense, or matter. Arguably, Husserl's talk of 'the *perceived as perceived*' can be understood as referring to an experience's purporting to be about something, and in some particular way. Thus, a deflationary view of intentional objects might give us some reason to adopt an understanding of noema as closely corresponding to the earlier notion of matter.

3.4 Sensation

I said in §3.1 that Husserl accuses Brentano of conflating the objects with the 'contents' of experience. This, suggests Husserl, is what leads Brentano to the implausible view that the object of an experience is a part of that experience itself. In a discussion clearly directed at Brentano's position, Husserl writes:

> These so-called immanent contents are therefore merely intended or intentional, while truly *immanent contents*, which belong to the real make-up of the intentional experiences, are *not intentional*: they constitute the act, provide necessary *points d'appui* which render possible an intention, but are not themselves intended, not the objects presented in the act. I do not see colour-sensations but coloured things, I do not hear tone-sensations but the singer's song, etc. etc.
>
> (1900–1, Vol. 2, p. 99)

Husserl's point here is that whilst Brentano is right to suppose that experiences have real, or immanent, 'contents', he is wrong to identify those as the objects of experience. That is, whilst there are such things as colour and tone sensations, these are not what we, respectively, see or hear. Such sensations, being part of the intrinsic make-up of an experience, are noetic. Further, since they are literally contained within – are parts of – experience, they may with some justification be referred to as 'contents'. It is crucial, however, not to confuse such 'immanent content' with the intentional content (i.e. matter) discussed in the previous section. It is sensible to use another word and Husserl himself provides us with more than one, for example 'hyletic data' and 'sensation' (1913, §85). Contemporary philosophy of mind provides us with another, 'qualia'. To my ear, 'sensation' is the most natural term to adopt for this purpose.

On Husserl's view, then, some intentional experiences have not only intentional features, but non-intentional, sensational features also. These non-intentional sensations are 'animated' by intentions which 'interpret' them. When, for example, I visually perceive a yellow banana, my yellow-ish visual sensations are 'animated' into an intentional experience of a banana. Husserl thinks of this animation as the giving of sense, or meaning (1913, §85). Taken on their own, sensations are not intentional, but they are parts – in Husserl's terminology, 'moments' – of experiences which also possess intentional features. As such, the acceptance of such non-intentional features does not pose a threat to the claim that all experience is intentional, although it does contradict the claim that all mental entities are intentional (taking 'entity' to range over both experiences and their features). Whether this is in tension with Brentano's Thesis that intentionality is the mark of the mental, then, will depend on exactly how that claim is formulated.

Why, in describing visual sensations, do I say 'yellow-ish' rather than yellow? Husserl insists that I do not *see* my visual sensations, rather I see objects themselves. But it is the objects we see that are yellow, not the experiences we have in seeing them. So if sensations are non-intentional features of experiences, they are not yellow, or any other colour. Nevertheless, as has frequently been recognised, there is an affinity between presented qualities, e.g. colours, and the putative sensations one has in being presented with those qualities (Peacocke, 1983, Ch. 1). In seeing a banana, I am aware of something yellow and my experience has a corresponding sensational property. Hence 'yellow-ish'.

The existence of sensations, conceived as non-intentional properties of experience, is controversial. As we shall see in the next chapter, for example, Heidegger's description of perceptual experience gives no role to sensation. More explicit in his rejection, Sartre emphatically insists that consciousness is 'empty', that it is pure intentionality. More recently, a number of philosophers have made similar claims, arguing that we have no need to appeal to such features of experience in order to give an adequate account of various forms of perception (e.g. Tye, 2002). In the following chapter we will look at this issue in some detail, assessing the case for sensation in perception.

4 Being-in-the-world

I mentioned in §1 that, according to Heidegger, the most fundamental form of directedness is practical. It is now time to elaborate on that passing remark and outline one of the most fundamental notions in Heidegger's philosophy, that of being-in-the-world. Heidegger's account of being-in-the-world can be seen as a critique, in certain respects at least, of the Husserlian account of intentionality. We need to look at what being-in-the-world is supposed to be; at what its connection is to those elements of Heidegger's conception of phenomenology introduced in the previous chapter; and at its relation to intentionality as described by Husserl. Being-in-the-world is something that will be important in later chapters, so it is crucial here to get a good sense of it and its place in the theory of intentionality.

Like Husserl, Heidegger takes intentionality to be a fundamental feature of experiencers such as ourselves. Employing his term *Dasein*, which literally translates as 'being there' but is typically used by Heidegger as equivalent to 'human being' or 'man', he writes that '[i]ntentionality belongs to the essence of the Dasein' (1927b, p. 157). Further, there are aspects of Husserl's character-isation of intentionality with which Heidegger is in agreement – he agrees, for example, that intentionality is not relational and, in particular, that it is not a relation to an immanent, or mind-dependent, entity (see the discussion of perception in the next chapter). Other aspects of Husserl's view, Heidegger arguably rejects – there is a case to be made, for example, that Heidegger's account of the role of social norms in intentionality involves a form of social externalism, and so the rejection of Husserl's apparent semantic internalism (for discussion see Wrathall, 1999 and Carman, 2003, Ch. 3). I will briefly return to this issue in Chapter 9. Of fundamental importance, however, is the fact that Heidegger considered Husserl's account of intentionality to be radically incomplete as an account of our directedness toward worldly entities or, as he puts it in *The Basic Problems of Phenomenology*, '*the sole characterisation of intentionality hitherto customary in phenomenology proves to be inadequate*' (1927b, p. 161). To see exactly what Heidegger finds wanting in the Husserlian conception, we must return to Heidegger's account of the pre-ontological understanding of being introduced in the previous chapter.

On Heidegger's view, for an entity to be accessible to me at all, I must possess an implicit understanding of its being. What this amounts to is a kind of know-how, an understanding of how to engage with entities of the sort in question. But this is only half of the story for, according to Heidegger, for something to be accessible to me, I must not only possess an implicit under-standing of its being, I must have an analogous understanding of my own being. And the reason for this is that my understanding of the being of entities is always of them as belonging to a world. They do not, however, belong to just any world, but to *my* world – the world that is partly constitutive of my own being. As Heidegger writes:

> Dasein exists in the manner of *being-in-the-world*, and this *basic determination of its existence* is the *presupposition for being able to apprehend anything at all*. By hyphenating the term we mean to indicate that this structure is a unitary one.
>
> (1927b, p. 164)

I have occasionally been using the term 'worldly' to refer to entities other than ourselves – the sort of thing that philosophers typically have in mind when they speak of 'the external world'. Heidegger has a different sense of world in mind. Fundamentally, *the world* in Heidegger's sense is a meaningful whole, a context in which individual entities gain their sense. It is helpful to think not of the philosophers' notion of 'the external world' but of the more colloquial use of world in such phrases as 'the world of higher education', or 'the world of the pop-star'. These 'worlds' are systems of norms and meaningful relations between entities of various sorts: in the world of higher education, an F means you're in trouble; in the world of the pop-star, image is everything; and so on.

In a brief discussion of Johann Gottlieb Fichte's famous instruction to 'Think the wall', Heidegger offers the following observation:

> If we are actually thinking the wall, what is already given beforehand, even if not apprehended thematically, is living room, drawing room, house. A specific functionality whole is *pre*-understood [...]. Existing in an environment, we dwell in such an intelligible functionality whole. We make our way throughout it. As we exist factically, we are always already in an environing world.
>
> (1927b, p. 64).

There is a great deal that can be said about this passage and surrounding discussion (1927b, §15; cf. the corresponding material in *Being and Time* entitled 'The Worldliness of the World'). We will return to the notion of the environment in Chapter 4, in Heidegger's account of 'the environmental thing'. The crucial point here is that in being intentionally directed to something ('Think the wall'), I am already aware of the context, or 'environing world', in which it is presented.

This, then, is the 'world' that we are 'being-in'. Each individual will have a slightly different such 'world': the various rooms and contents of my home, for example, mean something different to me than they do to you. And this is associated with the fact that the sense in which I am 'in' the world differs from the sense in which other entities are. I am not '*also* present in the lecture hall here, say, like the seats, desks, and blackboards' (1927b, p. 164), rather I 'dwell' in the world. I occupy a special position in my world because I am the one that *has* it (Heidegger addresses the objection that these considerations threaten to make the world subjective in 1927b, pp. 167–70).

Being-in-the-world is, according to Heidegger, a more fundamental form of directedness than intentionality as described by Husserl, although it must be noted that Husserl's introduction, in Part III of *The Crisis of European Sciences and Transcendental Phenomenology*, of the concept of the 'life-world' which is taken for granted in all thinking, is sometimes thought to be a response to Heidegger's critique (see the essays in Hyder and Rheinberger, 2009; also Føllesdal, 2000). It is a basic form of 'transcendence', in the sense that it involves us 'going beyond' ourselves into the 'world'. In being intentionally directed to an object, I am in each case already familiar with the world in which that object is situated and, since I am being-in-the-world, also with myself. And since this is an awareness of a unified phenomenon of Dasein and world, it is, on Heidegger's view, a form of self-awareness that is prior to the distinction between subject and object, a distinction that Husserl seemingly takes for granted (Heidegger, 1927b, pp. 158–61).

This familiarity with the world is, as mentioned, to be understood in practical terms. My familiar understanding of the environing world in which I am now situated (my home) involves, for example, knowing how to get from the kitchen to the bathroom, knowing how to lock the awkward front door, and so on. Since Ryle's (1949, Ch. 2) well-known account, it is standard to suppose that know-how is not reducible to knowledge-that (but for a recent attack on this orthodoxy, see Stanley and Williamson, 2001). As Ryle points out, for example, it seems to be 'possible for people to intelligently perform some sorts of operations when they are not yet able to consider any propositions enjoining how they should be performed' (1949, p. 31). That is, there are things that we can do – for example, tying shoelaces or riding a bicycle – without being able to describe how it is that we can do them. If being-in-the-world is, at least in part, a matter of understanding the being of entities by way of a capacity to practically engage with them, then perhaps it is something that we are not, ordinarily at least, in a position to formulate propositionally.

Dreyfus (1991) goes further than this, claiming that in his account of being-in-the-world, Heidegger has uncovered a form of intentionality that is both non-representational and non-mental. This is a form of intelligent behaviour that he calls 'absorbed coping'. Whilst we have not really addressed the issue of the relation between intentionality and representation, we may take Dreyfus's claim to involve at least the claim that absorbed coping is not intentional in any sense that Husserl's account would allow, but only in some broader sense of the term. In Dreyfus's words, 'Heidegger holds that all relations of mental states to their objects presuppose a more basic form of being-with-things which does not involve mental activity' (1991, p. 52). Again, '[his] description of the skilled use of equipment enables Heidegger to introduce [...] a new kind of intentionality (absorbed coping) which is not that of a mind with content directed towards objects' (1991, p. 69). On this picture, then, being-in-the-world is a form of intentionality – or directedness – that is fundamentally different in kind from that described by Husserl.

If Dreyfus is correct, then being-in-the-world constitutes a counterexample to Brentano's Thesis that intentionality is the mark of the mental, for being-in-the-world is certainly intentional. Dreyfus's account has, however, been the subject of much critical discussion. For example, Christensen (1997, 1998) argues forcefully that Dreyfus misinterprets Heidegger, whose account does not depart in this way from a broadly Husserlian conception of intentionality (also see McManus, 2012, Ch. 4). Blattner (1999b), on the other hand, defends key elements of Dreyfus's interpretation of Heidegger, arguing that both coping and perception can be non-representational.

Whether or not Heidegger thought of being-in-the-world as Dreyfus suggests, it is a view that can be evaluated on its own terms. Dreyfus, for one, finds it plausible, writing that

> [p]henomenological examination confirms that in a wide variety of situations human beings relate to the world in an organized purposive manner without the constant accompaniment of representational states that specify what the action is aimed at accomplishing. This is evident in skilled activity such as playing the piano or skiing.
>
> (1991, p. 93)

This goes beyond Ryle's claim that I may be unable to formulate *how* to do what I am doing, claiming that I am not representing even *what* I am doing. Even more radical than this, Dreyfus attributes to Heidegger and presumably endorses the claim that, 'when we carefully describe everyday ongoing coping activity we do not find any mental states' (1991, p. 86). Not only, then, are there no contentful states guiding and presenting the goal of my absorbed coping, there are no mental states involved at all.

This is likely to be disputed by many. Surely, when playing the piano I am aware of what I am doing – namely, playing the piano – even if, as Ryle points out, I am unable to say exactly *how*. Not only this, I can feel the keys under my fingers, see the sheet music, and so on. In short, it can be argued that there are very many intentional experiences (in Dreyfus's terms, 'mental states') that I have as part of the exercise of my capacity to play, which capacity is a manifestation of being-in-the-world and my pre-ontological understanding of being. This is not the place to investigate Dreyfus's position in detail, as much can be said on either side. One of the things at issue here is the relation between the sorts of unreflective action to which Dreyfus draws attention, and the varieties of perceptual experience. It is to perceptual experience that we turn in the next chapter.

5 Conclusion

Intentionality is a foundational feature of experience and absolutely central to the Phenomenological tradition. At issue is the most basic fact of experience: that objects are there for us. Whilst the status of intentional objects is at the heart of the apparent dispute between Brentano and Husserl, the claim of either account to describe the most fundamental way in which we are directed to the world and entities within it is contested by Heidegger. As we shall see, these issues are relevant to almost every phenomenological debate.

3 Experiencing things

in perception the perceived entity is bodily there

Heidegger, *History of the Concept of Time*

The nature of perceptual – and in particular visual – experience has, for a long time, been a central philosophical concern. In the natural attitude it is tempting to suppose that perceptual experience is a relation in which we stand to Objects: ordinary three-dimensional things such as people, trees, tables and chairs. To differentiate such entities from the properties and events that will be the focus of Chapters 4 and 5 respectively, I will refer to them as 'things'. The view that forms a part of the natural attitude, then, is that in enjoying a perceptual experience of a thing, I am related to an Object, an entity independent of me and my awareness of it. This is a version of the naïve realist view briefly mentioned in Chapter 1.

Despite the naturalness of this thought, however, it is a naïvety that many philosophers have been keen to displace. Consider, for example, the following passage from Hume:

> The table, which we see, seems to diminish, as we remove farther from it: but the real table, which exists independent of us, suffers no alteration: it was, therefore, nothing but its image, which was present to the mind [...] the existences, which we consider, when we say, *this house* and *that tree*, are nothing but perceptions in the mind, and fleeting copies or representations of other existences, which remain uniform and independent.
>
> (1739–40, p. 118)

Hume is here presenting an argument for what might now be called the sense data theory of visual experience (cf. Foster 2000, pp. 147–70), an argument which apparently depends, at least in part, on the phenomenological claim that that of which we are visually aware varies with our relation to it. This argument raises a challenge for the naïve conception of perceptual experience as a relation to Objects. More generally it raises the question of what it is that we are aware of in perceptual experience. It is that question, and a number of related issues, that we shall investigate in this chapter. As we will see, the previous discussion of intentionality will be crucial at a number of points.

Before we proceed, however, it should be pointed out that, as in the previous chapter, a number of the considerations in this chapter involve stepping outside the confines of the phenomenological reduction. For example, Hume's argument above rests, in part, on the claim that 'the real table, which exists independent of us, suffers no alteration'. This is an assertion that, on the face of it, presupposes the reality of Objects and so is, in that sense, non-phenomenological in the strict Husserlian sense. This becomes even more evident in the discussion of the argument from hallucination below. We will often be standing, then, with one foot in the natural attitude.

1 The case for sense-data

The view that, in our perceptual experience of things, we are directly aware not of Objects but rather of images or sense-data – mental entities that somehow represent Objects – should remind us of Brentano's doctrine of intentional inexistence. According to one interpretation of that view, the objects of intentional states are not Objects, but are themselves contained within the mind. Members of the Phenomenological tradition from Husserl onwards are united in rejecting this picture. As we have seen above, however, there do exist arguments for the view. We shall briefly consider three: the arguments from perspectival variation, from illusion, and from hallucination.

1.1 Perspectival variation

A fundamental feature of the phenomenology of visual experience is that it is perspectival. As I look through my window, I see a tree *from a particular vantage point*. As I move closer to it, it occupies a larger portion of my visual field; if I walk around it, its silhouette alters; and so on. According to Hume's argument, the fact that the tree appears to change in shape as I move in relation to it, alongside the obvious fact that the actual tree does not in fact so change, entails that it is not the tree of which I am aware but rather an 'image' or 'perception'.

The Scottish empiricist philosopher Thomas Reid (1710–1796) famously objected to Hume's argument, pointing out that the fact that the table, or tree, *appears* to change in shape and size is consistent with the fact that it *does not* so change. Reid, although accepting both premises, thus refuses to draw Hume's conclusion. As he says, 'apparent magnitude is the middle term in the first premise; real magnitude in the second' (Reid, 1786, p. 178). Hume's argument, it seems, presupposes that the objects of visual experience, whatever they are, must have just those properties that they appear to have. This is a claim that has, following Robinson (1994, p. 32), become known as the Phenomenal Principle. Only if this principle were true would Hume's first premise be inconsistent with the fact that the table, or tree, is perceived yet does not itself change as one moves around it. Reid's response can be seen as asserting that this presupposition is not mandatory.

A slightly different line of thought runs through Husserl's influential account of perceptual variation. He writes:

> Constantly seeing this table and meanwhile walking around it, changing my position in space in whatever way, I have continually the consciousness of this one identical table as factually existing 'in person' and remaining quite unchanged. The table-perception, however, is a continually changing one; it is a continuity of changing perceptions.
>
> (1913, p. 86)

Husserl here claims that although my experience alters as I walk around a thing, the thing does not. Indeed, he further claims that the thing is itself *presented as* unchanging. The thing not only remains constant, it appears to. Notice that this is a claim that Husserl can make from within the confines of the phenomenological reduction. It concerns how the object is presented. Crucial here is the Husserlian notion of an 'adumbration': a way in which a thing's features appear. Objects, according to Husserl, are given 'in' adumbrations. Since every visual experience is from somewhere, visually perceived things are always presented from some particular angle and this affects the way they look. Some features are occluded and those that are available to vision vary with viewing angle. As Husserl says, '[o]ne and the same shape (given "in person" *as* the same) appears continuously but always "in a different manner," always in different adumbrations of shape' (1913, p. 87). The object, and its shape, remains constant despite being given 'in' varying adumbrations.

With his notion of adumbration, Husserl does acknowledge perspectival variation. He denies, however, that the Humean conclusion can be drawn. For whilst adumbrations vary, it is not they, but the Object itself, that is the object of perceptual experience. As Heidegger makes the point, 'it is not an *adumbration* which is intended, but the perceived thing itself, in each case in an adumbration' (1925, p. 43). One way to put this would be to say, in relation to the example with which we began, that Hume is wrong to claim that things look smaller as we move away from them. Rather, their apparent size remains constant, it is *how that size is adumbrated* that varies. We shall return to the Husserlian notion of adumbration both later in this chapter and, in more detail, in the discussion of perceptual constancy in Chapter 4.

1.2 Illusion

Consider next, cases of perceptual illusion, what Heidegger calls 'deceptive perception'. He gives us the case in which 'I am walking in a dark forest and see a man coming toward me; but upon closer inspection it turns out to be a tree' (1925, p. 30). Or consider a simpler case in which a straight stick placed in a glass of water appears bent. In each example, we are aware of something – tree and stick respectively – that appears to be some way other than it is. The argument from illusion moves from the claim that in, for example, the latter case we are visually aware of something bent, and the fact that the stick is not bent, to the conclusion that we are not visually aware of the stick itself but of a sense-datum that really is bent.

There is obviously a similarity between this and Hume's argument from perceptual variation. In fact, on the Humean formulation, according to which things appear to get smaller as one moves away from them, that argument is a version of the argument from illusion, since presumably things don't get smaller when one moves away. Notably, Husserl's description of the phenomenon of perspectival variation does not have that consequence. We shall consider this issue in more detail in Chapter 4. Here we can simply note that, just like the argument from perspectival variation, the argument from illusion seemingly depends on the Phenomenal Principle; the assumption that if something visually appears *F* to one, then one is visually aware of something that is *F*. As a result, if one is willing to deny the assumption, as we have seen are both Reid and Husserl, then the argument will seem to have little to commend it.

1.3 Hallucination

Perhaps the most powerful case for sense-data is the argument from hallucination. Heidegger, addressing his students, suggests that

> [i]t is possible for my psychic process to be beset by a hallucination such that I now perceive an automobile being driven through the room over your heads. In this case no real object corresponds to the psychic process in the subject.
>
> (1925, p. 30)

Let us suppose that Heidegger's hallucination is so realistic as to be completely indistinguishable, to him, from a corresponding non-hallucinatory experience. That is, it is, for Heidegger, *just as if* a car really is driving through the room. The argument can be stated like this: when Heidegger *hallucinates* an automobile, although he is aware of something that appears to be an automobile, he is not aware of an automobile. But, since he has an experience that is indistinguishable from the experience of actually *seeing* an automobile, he is in the very same experiential state in both cases. So, when he really does see an automobile, he is not – at least not *directly* – aware of an automobile. Rather, in both cases, he is directly aware of something mental – image, sense-datum, etc. – which *represents* an automobile. At best, then, visual experience gives us only indirect awareness of things.

This argument does not rely on the assumption that if something visually appears *F* to one, then one is visually aware of something that is *F*. It is not, then, to be dismissed as easily as the arguments from perspectival variation and illusion. The argument from hallucination does, however, presuppose that in hallucination, one is aware of *something*; that when we hallucinate we are related to something. This follows from one of the assumptions of the naïve view of perceptual experience – that it is a relation to something – alongside the entirely plausible thought that hallucinations are perceptual experiences. One way to challenge the argument from hallucination, then, is to deny that perceptual experience is a relation. This is the route taken by those endorsing an intentional account of perceptual experience.

2 Perceptual experience as intentional

In the 'Preliminary Part' of *History of the Concept of Time* Heidegger articulates a broadly Husserlian account of the main features of phenomenology. During his discussion of intentionality, Heidegger addresses the issue of perceptual experience, telling us that intentionality, that fundamental characteristic of all experience, is not relational, but intrinsic:

> It is not the case that a perception first becomes intentional by having something physical enter into relation with the psychic, and that it would no longer be intentional if this reality did not exist. It is rather the case that perception, correct or deceptive, is in itself intentional. Intentionality is not a property which would accrue to perception and belong to it in certain instances. As perception, it is *intrinsically intentional*, regardless of whether the perceived is in reality on hand or not.
>
> (1925, p. 31)

This is evidently an application, to the case of perceptual experience, of the more general claim that intentionality is not a relation to Objects (also see the discussion of perception in Heidegger, 1927b, §9 (b)). In the previous chapter we saw both Brentano and Husserl suggesting, despite sometimes speaking of it as such, that intentionality is not a relation to Objects, but rather an intrinsic feature of intentional experiences themselves. If, then, perceptual experience is intentional, or as Heidegger puts it, '*perceiving is a directing itself towards*' (1925, p. 32), it would follow that perceptual experience is not relational; that the assumption common to both the naïve view of perceptual experience as a relation to Objects and the Humean view of perceptual experience as a relation to sense-data, is false.

What, then, are we to make of the possibility of hallucination? On the intentional view, it is wrong to claim that there is something – some thing – of which the hallucinating person is aware. On this view, rather, the possibility of hallucination is accounted for by the fact that my perceptual intentions can fail to pick out an object, they can be 'non-veridical'. When Heidegger hallucinates an automobile, he perceptually intends an automobile, but there is none. Hallucinatory perceptual experiences have an intentional object that is not an Object, i.e. is not a mind-independent thing. We can think, then, of the possibility of hallucination as showing that perceptual experiences do not depend on the existence of Objects. The sense-data view, however, wrongly supposing that perceptual experience must be a relation to *something*, invents subjective entities for the purpose. The view of perceptual experience as intentional will reject this last move, denying that perceptual experience is a relation at all.

The intentional view of perceptual experience, then, has a neat account of hallucination. Something similar can be said for illusion. If a hallucination is a case of perceptually intending an object that does not exist, an illusion is a case of perceptually intending an object as being a way that it is not. Recall that on the Husserlian account of intentionality, intentional experiences have both quality and matter. The quality, in the present case, is *visual perception*. The matter, on the other hand, determines both the object and the way in which it is presented. In the case of perceptual illusion, we may say that the object (the stick) is presented in a way (as bent) that it is in fact not.

Heidegger presents a number of considerations that tell against the sense-data (image) view and speak in favour of the intentional account. In both respects he cites the phenomenology of visual experience as decisive. For example, in the case of hallucination, he asks:

> is not the hallucination in its own right a hallucination, a presumed perception of an automobile? Is it not also the case that this presumed perception, which is without real relationship to a real object, precisely as such is a directing-itself-toward something presumably perceived? Is not the deception itself as such a directing-itself-toward, even if the real object is in fact not there?
>
> (1925, p. 31)

That is, the fact that hallucinations do not manage to pick out Objects does not alter the fact that, to the one undergoing a hallucination, it *seems* to. The object of visual experience is given as an Object in cases of veridical perception and hallucination alike. Hallucination is characterised precisely as occurring when, although there is no Object to which one is perceptually related, there *seems to be*. So while the possibility of hallucination shows that perceptual experience is not a relation to Objects, the intentionality of perception explains why it *seems* to be. Thus, the intentionality of perception itself explains why the natural attitude's naïve picture of perceptual experience is so natural.

Heidegger's central objection to the sense-data (image) view is similarly based on the claim that it is simply untrue to the phenomenology of perceptual experience itself. As he claims:

apprehension of a picture, the apprehension of something as something pictured through a picture-thing, has a structure totally different from that of a direct perception. [...] In the consciousness of a picture, there is the picture-thing and the pictured. [...] In simple perception, by contrast, in the simple apprehension of an object, nothing like a consciousness of a picture can be found. It goes against all the plain and simple findings about the simple apprehension of an object to interpret them as if I first perceive a picture in my consciousness when I see that house there, as if a picture thing were first given and thereupon apprehended as picturing that house out there.

(1925, p. 42)

We will return to the experience of pictures in Chapter 6. Here, notice that, as Heidegger describes it, in the paradigm instances of awareness of a picture, we are aware of (at least) two things. We are aware of the 'picture-thing' – the paper, screen, etc. – and *through* this we are aware of that which is pictured. Further, the way in which a pictured thing is presented differs from the way in which a perceived thing is. In particular, a pictured thing is not, whilst a perceived thing is, as both Husserl and Heidegger put it, 'bodily given' (*Leibhaftigheit*). Thus, the sense-data view would entail that, in perceptual experience, sense-data (images) are bodily given and the Objects they represent are given in the way that pictured things are. But this is simply false. On the contrary, in perceptual experience it is Objects themselves that are bodily given or, in the case of hallucination, seem to be. Thus, on Heidegger's assumption that the view ought to be modelled on paradigm instances of picture perception (in which a picture is bodily given and a pictured thing is given in some other way), the sense-data view is ruled out.

3 Bodily presence

Husserl and Heidegger claim that in perception we encounter things as 'bodily present' (also, 'in the flesh', 'bodily given', or 'bodily there'). In each case the idea is that this feature distinguishes the way in which perceived objects are given from the way in which, for example, merely judged things are given. Things are given *via* judgement but they are not experienced as being bodily present.

3.1 Bodily presence and empty intending

Bodily presence has a crucial role to play in intentional accounts of perception such as are defended by both Husserl and Heidegger. According to intentional views, perceptual experience involves the awareness of ordinary things presented as being some way or other. But exactly the same can be said of judgement. I may have a visual experience of the tree as tall, or I may judge the tree to be tall. Regarding this, and employing Husserl's terminology from *Logical Investigations*, one might suppose that the intentional experiences in question have the same object (the tree), and the same intentional matter (that it is tall). But the two experiences are, of course, phenomeno-logically distinct. Visually perceiving a tree as tall is entirely different from judging it to be so. How, then, are we to mark this phenomenological difference in our account? How can we distinguish perceptual experience from other intentional experiences such as experiences of judging?

Of course, the two experiences will be said to possess a different intentional quality: one is perceptual and the other judgemental. But without further elaboration this is just a label, giving us little insight into what the difference really amounts to. The same complaint can be levelled at the bare assertion that in perceptual experience, but not judgement, things are given as bodily present. For we will rightly demand to know in what bodily presence consists.

Some progress is made with the notion of 'intuitive fulfilment'. In perception, things are given *intuitively*; my perceptual experience is intuitively fulfilled. In judgement, by contrast, things are given *non-intuitively*; my experience is an 'empty intending' (Husserl, 1900–1, Vol. 2, pp. 233–5; cf. Heidegger, 1925, p. 41). We can get an initial grip on this notion by looking a little more closely at the visual perception of a thing. For, at least according to Husserl and, following him, Heidegger, the visual perception of things involves both intuitively fulfilled and intuitively empty aspects.

3.2 Intuitive fulfilment and co-presence

Let us return to the fact that perceptual experience is perspectival. The fact that I am looking at the tree from a particular location relative to it means that some of its features, being opaque, occlude others. To put it somewhat tenden-tiously, I see only one side of it. As Husserl puts it, '[o]f necessity a physical thing can be given only "one-sidedly"' (1913, p. 94). This is, of course, related to the idea that perceived things are given *via* adumbrations. We shall return to the question of exactly what an adumbration is. For now, notice that, despite the fact that my visual experience only presents one side of (so, a part of) the tree, it seems to be false to say that what I am visually aware of is anything less than a tree. I do not see a tree-side or, perhaps more accurately, a tree-facing-surface. Rather, I see a tree.

This is not just true of how we speak of the objects of visual perception, but also of the phenomenology of visual experience itself. What seems to be given to me in vision is a three-dimensional solid thing, complete with occluded aspects. But this is apt to seem puzzling. How is it that, on the one hand, I am presented with only one side of the tree – and so less than a whole tree – yet, on the other hand, I am visually aware of nothing less than the whole tree?

This air of paradox is somewhat dispelled with the introduction of the notion of intuitive fulfilment. As Husserl describes it, every perception of a thing involves the intuitively full presentation of some aspects of that thing and the intuitively empty presentation of others. That is, whilst the tree as a whole is visually present, only that part of it that faces me is so with intuitive fullness. So,

> Of necessity a physical thing can be given only 'one-sidedly;' [...] A physical thing is necessarily given in mere 'modes of appearance' in which necessarily a *core of 'what is actually presented'* is apprehended as being surrounded by a horizon of *'co-givenness'*.
>
> (1913, p. 94)

The facing side of a seen object is 'actually presented' – it is given with intuitive fullness. The rear side is 'co-given' – it is given in an empty way. Thus, the solution involves the claim that there are (at least) two distinct ways in which a thing's aspects can be present to vision. The side of the tree that I am facing is visually present to the fullest extent. The side of the tree that I am not facing, whilst still (visually) appearing to be there – after all, the tree looks three, not two, dimensional – does so in a less than full way, it is co-presented ('co-given'). That which is merely co-given is a 'horizon', of which there are two sorts: internal and external (Husserl, 1954, §47). The internal horizon in an experience of some thing includes those aspects of the object (rear aspect and insides) that are co-given. The external horizon includes other things, co-given as part of the surrounding environment – the background against which the tree is perceived.

Husserl often uses the term 'anticipation' to describe the way in which the merely co-presented is given in perceptual experience. This, furthermore, is understood in terms of other possibilities of perception:

> there belongs to every external perception its reference from the 'genuinely perceived' sides of the object of perception to the sides 'also meant' – not yet perceived, but only anticipated and, at first, with a non-intuitional emptiness [...] the perception has horizons made up of other possibilities of perception, as perceptions that we *could* have, if we *actively directed* the course of perception otherwise.
>
> (1931, p. 44)

In these terms, whilst only the front aspect of an object is 'genuinely perceived', its other features are also visually present by way of being anticipated. This anticipation consists, at least in part, in perceptual expectations as to how the object will appear in subsequent experiences. Further, whilst these anticipations count as genuinely perceptual, they lack the intuitional fullness of the facing side of the perceived thing.

Those aspects of a thing that are merely co-given can be brought into intuitional fullness precisely by moving around the thing, or by the thing's itself rotating. This gives us a clue as to the structure of perceptual anticipations. They take the form of conditionals associating potential move-ments with further perspectival views of the thing. And, of course, not just any old perspectival view will be anticipated but they will follow a law-like pattern dictated by those properties the thing seems to have. Thus, for example, I will perceptually anticipate that if I move towards some visually perceived thing, it will be presented through adumbrations that take up an increasing proportion of my visual field (as we might put it, they will become larger-ish). The perception of something as a thing involves this interplay between those aspects of it that are given with intuitive fullness and those aspects that are co-given *via* anticipation.

A great many questions arise concerning this phenomenological description of the perspectival nature of the visual experience of things, some of which we will address in the next chapter. Our purpose here is simply to introduce the notion of intuitive fullness with the express aim of clarifying the difference between perception and judgement. Perceptual experience involves things being given as bodily present. Things given as bodily present are, at least in part, given with intuitive fullness. Judgement lacks all of this. When I judge that the tree is tall, the tree itself is 'before my mind', but not with the intuitive fullness (and associated anticipation) characteristic of the presentation of the facing side of the visually perceived tree.

3.3 Bodily presence and sensation

We have, with the difference between intuitive fullness and emptiness, an initial grasp on the notion of bodily presence. We can, however, dig a little deeper and ask what it is in virtue of which the intuitively full is *intuitive*. One seemingly plausible answer to this question, and the one given by Husserl, is that the intuitive is that which involves *sensation*.

As we saw in the previous chapter, Husserl claims that intentional experiences have, in addition to their intentional features, non-intentional, sensational features. When, for example, I visually perceive a yellow banana, my yellow-ish visual sensations are 'animated' into an intentional experience of a banana. To this we can now add that, on Husserl's view, examples of such sensational features of experience are the adumbrations mentioned above. As the lighting conditions change, the constant colour of the banana will be given *via* varying yellow-ish adumbrations. As I walk around the tree, its constant shape will be given in varying shape adumbrations. Husserl writes:

> each phase of the perception necessarily contains a determined content of adumbrations of colour, adumbrations of shape, etc. They are included among '*the Data of sensations*,' [...] the Data are animated by '*construings*' within the concrete unity of the perception and in the animation exercise the '*presentative function*,' or as united with the construings which animate them, they make up what we call '*appearings of*' colour, shape, and so forth. These

moments, combined with further characteristics, are the really inherent components making up the perception.

(1913, p. 88)

Perceptual experiences have non-intentional, sensational features – adumbrations of colour, shape, and so on – and it is these that constitute their intuitive character. So, one difference between that which is given with intuitive fullness and that which is not, concerns the presence or otherwise of sensation. Whist the facing side of the tree that I see is presented by way of animated sensations, the rear side is not. That the tree is bodily given means, on Husserl's picture, that it is given with the peculiar combination of the intuitively full and empty. This, in turn, is understood, at least in part, in terms of the sensational features of perceptual experience – features lacking in experiences of judging.

This explanation of bodily presence in terms of sensation is one that will be challenged by many philosophers. One common concern is with the terms – such as 'yellow-ish' – with which one can describe perceptual sensations. For it is not obvious what it is for something to be yellow-ish or circular-ish (see Hopp, 2008 and 2011). One thing that is certain is that a yellow-ish sensation is not yellow, a circular-ish sensation is not circular, and so on. Sensations are not literally shaped or coloured. Only things are shaped or coloured. Husserl sometimes maintains that sensations *resemble* the properties in question:

> Only those contents can be intuitively representative of an object that resemble or are like it [...] we are not wholly free to interpret a content *as* this or *as* that [...] since the content to be interpreted sets limits through a certain sphere of similarity and exact likeness.
>
> (1900–1, Vol. 2, p. 244)

But it is difficult to see how to make sense of the idea that a sensation, itself lacking in colour and shape properties, could resemble yellowness or circularity. With this problem before us, perceptual sensations of the sort to which the Husserlian appeals can begin to look rather mysterious.

In the contemporary debate, a certain form of representationalism, typified by the work of Tye (2002), asserts that visual experience is exhaustively accounted for in terms of representational content or, to use Husserlian language, the experience's intentional matter. So, a representationalist of this sort will deny Husserl's claim that intentional experiences have, in addition to their intentional features, non-intentional, sensational, features. The thought here is that, in the case of seeing a yellow banana, there is no yellow-ish sensation, simply an awareness of the banana as a particular shade of yellow.

It is not entirely anachronistic to attribute such a view, or at least something very much like it, to Sartre who, although he accepts the broad outline of Husserl's view of intentionality, claims that all intentional experiences are empty. In an early essay he writes that '[t]here is nothing in it [consciousness] but a movement of fleeing itself, a sliding beyond itself [...] consciousness has no "inside'" (Sartre, 1939a, p. 4). Later, in the introduction to *Being and Nothingness*, Sartre is more explicit, claiming that Husserlian *hyle* (sensation) is unintelligible. As we have seen, these sensations are supposed to be, for example, red-ish but not red, resembling coloured things without themselves being coloured. It is easy to be sympathetic to the complaint of unintelligibility (but see Williford, 2013, for a defence of the notion). Perceptual experience, according to Sartre, is in reality nothing but directedness towards its objects. Reflection on it reveals it to be empty, lacking in the sort of non-intentional features to which Husserl appeals.

Heidegger's discussion of perceptual experience may be interpreted in a similar fashion. We can, at the very least, say that the concept of sensation plays no significant role in his account of intentionality. Whilst Heidegger does employ the notion of bodily presence, and also that of adumbration, these are not given a further analysis in terms of the sensational properties of intentional experiences. Heidegger's discussion of adumbration, for example, occurs within his account of 'the perceived of perceiving', in particular the *manner of its being intended* (noema), rather than of the perceiving itself (noesis). Heidegger writes:

> When I go into a room and see a cupboard, I do not see the door of the cupboard or a mere surface. Rather the very sense of perception implies that I see the cupboard. When I walk around it I always have new aspects. But in each moment I am intent, in the sense of natural intending, upon seeing the cupboard itself and not just an aspect of it. The thing *adumbrates*, shades off in its aspects. But it is not an *adumbration* which is intended, but the perceived thing itself, in each case in an adumbration.
>
> (1925, p. 43)

The implicit suggestion here is that adumbrations, varying as I move with respect to some thing, are 'aspects' of the thing I perceive rather than features of my experience of it.

The view that the phenomenology of the perceptual experience of things can be accounted for entirely in terms of how its objects are presented as being – that there is no need to appeal to non-intentional, sensational features of experience – can be challenged on a number of grounds. We can distinguish between two broad categories of problem case, each of which contrasts a pair of experiences. First, there are those in which, as the Husserlian would put it, we have the same sensational features but different intentional features. Second, there are those in which we have the same intentional features but different sensational features.

An example of the first kind is the familiar Necker cube. As is well known, Figure 3.1 can be viewed in two ways. The face ABCD can be seen as either to the front or the rear of the cube. How one sees the cube is, to a certain

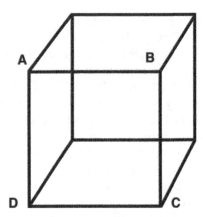

Figure 3.1 The Necker cube

extent, under one's own control. One can, pretty much at will, perform an 'aspect switch'. For our purposes, the interesting feature of such switches is that, when we compare the pre-switch experience (with ABCD at the front) with the post-switch experience (with ABCD at the rear), they seem to be in one way different and in another way the same. The difference consists in the cube's seeming to 'face' in a different direction. The sameness consists in the location of, and relations between, the various points and lines. The aspect switch seems to consist in altering the apparent direction of the cube without changing the apparent position of any of its parts.

The Husserlian accounts for this seemingly paradoxical state of affairs by distinguishing two features of visual experience. What remains constant through the aspect switch are the sensational features of the experience; what changes are the intentional features. As Husserl would put it, the sensations (*hyletic* data) are 'animated' by a different intentional matter, thereby constituting a new intentional object. This description is, of course, not available to those who wish to do without perceptual sensations. The challenge, then, for such a view is to describe the similarity between the pre-switch and post-switch experiences in purely intentional terms.

As an example of the second kind of case, we can consider something already encountered, and to be treated in more detail in the next chapter: size constancy. A nice illustration of this phenomenon is offered by Peacocke:

> Suppose you are standing on a road which stretches from you in a straight line to the horizon. There are two trees at the roadside, one a hundred yards from you, the other two hundred. Your experience represents these objects as being of the same physical height and other dimensions [...]. Yet there is also some sense in which the nearer tree occupies more of your visual field than the more distant tree.

(1983, p. 12)

As with the case of the Necker cube, we have here two experiences – or, more accurately, two parts of one complex visual experience – that seem to be in some way similar and in another way different. The experience of the nearer tree is similar to the experience of the farther tree in that both are presented as being the same size. The experiences differ in that the nearer tree seems to, as Peacocke puts it, 'occupy more of your visual field'. So, whilst they seem to be the same size, they appear somehow different with respect to size!

Once more, the Husserlian has an account of this seeming paradox, one that employs the distinction between the intentional and the sensational features of visual experience. In this case, what remains constant between the experiences of the trees is that they are intended as being the same size. What differs between the experiences is that each is associated with a qualitatively different visual sensation. As Husserl might put it, one and the same size is, in the two cases, differently adumbrated. As before, this description is not available to those who deny the existence of perceptual sensations. The challenge, again, for such a view is to describe the difference between the experiences of the near and far trees in purely intentional terms. We shall return to issues arising from the phenomenon of perceptual constancy in Chapter 4.

3.4 Bodily presence and non-conceptual content

Even if the above challenges can be met and the various problem cases be accounted for in purely intentional terms, there remains the question of how such a view would account for the difference between a visual experience of a tree as tall and an experience of judging the tree to be tall. As both Husserl and Heidegger insist, of the two only the perceptual experience presents the tree in the flesh, as bodily present. This, in turn, is further analysed in terms of intuitive fulfilment: only the visual experience possesses that characteristic combination of the intuitively fulfilled and the intuitively empty that we described above. Husserl goes one step further, analysing intuitive fulfilment in terms of perceptual sensations: visual experiences, but not judgements, have non-intentional, sensational features. Sartre, and arguably also Heidegger, decline to take this step. But if we do not endorse perceptual sensations, the question arises as to how we are to characterise the *intuitive* nature of perceptual experience.

A popular contemporary response to this question is to claim that whilst there is a sense in which the visual experience and the judgement share representational content (matter), there is a fundamental way in which their contents differ. This difference lies in the fact that whilst the judgement possesses only *conceptual* content, the visual experience possesses content that is *non-conceptual*. Non-conceptual content has been defined in a number of different ways, but it is relatively uncontroversial to say that content is non-conceptual if a person can have an experience with that content even though they lack the concepts required to articulate it. An example that illustrates this, and is often thought to motivate the claim that visual experience has non-conceptual content (Peacocke, 1992), is the fineness of grain of colour experience. As I look at the tree beyond my window, its leaves look a variety of shades of green; many more shades than I have names for. That is, I am unable to conceptualise the content of my visual experience of the tree; it has non-conceptual content.

If the term 'content' is allowed to include both intentional and non-intentional features of experience, then we might think of the Husserlian defence of visual sensation as a version of the claim that visual experience has some non-conceptual content. In contemporary discussions, however, it is more common to use 'content' to refer to the intentional matter of an experience (or its noema). On this understanding, the claim that visual experience has non-conceptual content is a claim about the specific way that visual experience presents things in the world to be. On this understanding, non-conceptual content is an intentional feature of experience.

If perceptual experiences possess non-conceptual content in this sense, and if experiences of judging do not, then it may be that we can distinguish between the two without appealing to non-intentional perceptual sensations. A number of concerns have been levelled against such an approach, however. I shall mention two. First, not only are there clear phenomenological differences between perceptual experiences and experiences of judging, there are also manifest differences between experiences in the various sensory modalities and it is not clear how the proposal deals with those. Second, it has been argued, most forcibly by McDowell (1994), that such a view falls into the myth of the given and should, for that reason, be rejected.

Taking the first worry first, suppose that I see a child's building block before me. On the present view, my visual experience will present it as, amongst other things, cubic. Now suppose that rather than see the block, I hold it in my hand, running my fingers over its sides and edges. We might be tempted to say that my tactile experience presents it as cubic. But now a question arises: if both the visual and the tactile experience represent the block as cubic, in what does the difference between the two experiences lie? A visual experience of a cubic block is very different, from the perspective of the experiencer, from a tactile experience of a cubic block.

One way to describe this difference would be to say that whilst they share intentional content, the two experiences differ with respect to the sensations involved. The first experience involves visual shape sensations, the second experience involves tactile shape sensations, and these are just primitively distinct. Of course, if one refuses to countenance perceptual sensation, this description will not be available. Nor is it possible to say, as one might when faced with the distinction between perception and judgement, that only one of the experiences in question possesses non-conceptual content. For, in this case, both experiences are perceptual, so both need to be distinguished from judgement, so both are equally plausible candidates for being non-conceptual.

The challenge, then, will be to describe the difference between visual and tactile shape experience in purely intentional terms.

Putting this issue to one side, we can consider the charge that the proponent of non-conceptual content must endorse the myth of the given. We encountered the myth of the given in Chapter 1. There we were concerned with Husserl's notion of eidetic intuition, the claim that phenomenology can gain insight into the essential nature of experience by way of describing intuited, or 'seen', essences. The challenge there was that it is a myth to suppose that the given can justify any one categorisation. Rather, to justify a perceptual judgement, the content of perceptual experience must itself be within 'the space of reasons'.

The proponent of non-conceptual content might be seen to be open to this charge. Non-conceptual content is, supposedly, content the specification of which requires certain concepts, but which is not itself conceptual. Thus, if perceptual experience is non-conceptual, then forming a (conceptual) perceptual judgement on its basis involves moving from non-conceptual to conceptual content. More than this, the non-conceptual content requires, and so legitimises, one particular conceptualisation over another. For example, if my visual experience non-conceptually presents a block as cubic, then this very content legitimises – provides a reason for – the judgement that the block is cubic, but not the judgement that the block is spherical. This is, almost by definition, the idea of the given. And as McDowell puts the problem, 'we cannot really understand the relations in virtue of which a judgement is warranted except as relations within the space of concepts' (1994, p. 7). That is, if perceptual experience has non-conceptual content, it makes no sense to suppose that it provides a *reason* for some particular perceptual judgement over another. For something to provide a reason, according to McDowell, it must be conceptually articulated. Lacking that conceptual articulation, all we can say is that perceptual experience *causes* perceptual judgement; we cannot say that it justifies it. McDowell memorably sums this line of thought up by claiming that 'the idea of the Given offers exculpations where we wanted justifications' (1994, p. 8).

The Husserlian position, according to which perceptual experiences have both intentional and sensational features, is in principle able to avoid this challenge. For it is open to proponents of the view to claim, first, that intentional content (matter) has *conceptual* content and, second, that the non-intentional (non-conceptual) features of experience do not legitimise, or provide reasons for, the intentional features that 'animate' them. On such an elaboration of the view, perceptual experience would, by McDowell's lights, be within the 'space of reasons'. Having said this, there is some reason to suppose that this elaboration would not accurately represent Husserl's own position and that, in fact, he does subscribe to non-conceptual content and is thus open to the charge of the myth of the given (Mulligan, 1995). If so, then this is an objection that Husserl must himself answer.

Of course, the claim that the given is indeed a myth is highly contentious. Opponents can point, on the one hand, to the conceptually unarticulable fineness of grain mentioned above and, on the other, to the fact that the denial of non-conceptual content appears to rule out the possibility that the conceptually unsophisticated (i.e. babies and animals) enjoy perceptual experience. And this is a claim that most will find unpalatable.

None of this, however, explains how it is that non-conceptual experience could justify conceptually articulated judgement. One suggestion, influentially made by Dreyfus (2005, 2013), is that such an account can be found in the Heideggerian notion of being-in-the-world, introduced in the previous chapter and, in particular, in the phenomenon that he calls 'absorbed coping'. The idea, in brief, is that through our bodily familiarity with our surroundings, we are aware of the world as 'affording' or 'soliciting' certain actions, 'our non-conceptual coping skills [...] disclose a space in which things can then be as what and how they are' (2013, p. 21). As I type, for example, my fingers 'cope' with the keys, engaging with them as keys, but not by means of conceptual representation. In this way, Dreyfus suggests, we can see that the given is no myth and that our engagement with the world can be non-conceptual yet nevertheless ground judgement since absorbed coping non-conceptually presents things 'as what they are'.

The debate between conceptualists and non-conceptualists remains lively (see, for example, the essays in Gunther, 2003), and the plausibility of Dreyfus's response to McDowell has been subject to much discussion (Schear, 2013). We will return to the notion of being-in-the-world at a number of points in subsequent chapters. For now, the central point is that whilst the Husserlian account of bodily presence in terms of perceptual sensations can, at least in principle, sidestep the issue of the myth of the given, the alternative account of bodily presence in terms of non-conceptual content cannot. It must meet the challenge head on.

4 Conclusion

Perceptual experience would appear to be the simplest and most fundamental way in which we are in touch with the world around us. Intentional accounts of perception walk a line between the objectivism of naïve realism and the subjectivism of sense-data theory. But the question of how perceptual experience can be distinguished from judgement, and how the senses may be distinguished from each other, raises deep questions about the relation between intentionality and sensation. What has also come into view is that not only must the phenomenologist describe the experience of things, an account is also needed of how we experience their properties: their shapes, sizes, and colours.

4 Experiencing properties

Nothing is more difficult than knowing precisely *what we see*
Merleau-Ponty, *Phenomenology of Perception*

In perceptual experience I am aware not only of things but also of their properties. As I look at the tree beyond my window I am aware of its irregular shape, its size, the colour of its leaves and bark. As Husserl emphasises, when I change position with respect to some thing, although my experiences change, the object looks to be *the same*, and it does so in two ways. Not only does the object appear to remain the same thing through such changes, it also appears to *have the same properties*. This is, puzzlingly, despite the fact that those properties appear differently. In this chapter we will consider a number of ways in which this puzzle has been approached, with a focus on Husserl and Merleau-Ponty. As we shall see, the discussion will draw on that of the experience of things in the previous chapter. It will also, with the account of Merleau-Ponty, look forward to the discussion of the experience of the body in Chapter 8. This will lead us to a discussion of Heidegger's claim – indicated in the earlier discussion of being-in-the-world – from *History of the Concept of Time*, and later elaborated in *Being and Time*, that what we perceive is given first and foremost as an *environmental* thing.

1 Perceptual constancy

1.1 The experience of shape, size, and colour

Consider the way in which the shapes, sizes, and colours of things are given in visual experience:

Shape: There is a sense in which a thing looks to be the same shape despite the fact that it is seen from different angles. That is, the apparent shape of a thing is not determined by the shape that you would have to draw on paper to accurately portray the way it looks. But despite the fact that things look the same shape from different angles, there is also a clear sense in which they look different 'shape-wise'. But, of course, objects do not look to have two incompatible shapes! How can this be?

Size: There is a sense in which a thing looks to be the same size, despite the fact that it 'ought' to look larger close up and smaller far off. That is, the apparent size of a thing is not determined by the area of the visual field that it takes up. But despite the fact that things look the same size from different distances, there is also a clear sense in which they look different 'size-wise'. But, again, objects do not look to have two incompatible sizes!

Colour: There is a sense in which the colour of a thing appears to be the same despite the fact that it 'ought' to look darker in the shade and lighter in sunlight. That is, the apparent colour of the thing is not determined by the colour of paint that one would require to accurately depict it. Once more, despite the fact that objects look the same colour in different lighting conditions, there is a clear sense in which they look different 'colour-wise'. Again, how can this be, given that objects certainly do not look to have two incompatible colours?

This is the phenomenon of perceptual constancy which, as can be seen, presents a puzzle about the way that things appear. Whilst the above examples all concern vision, similar phenomena occur in other modalities. Consider, for example, how the volume of the noise made by a pneumatic drill alters yet stays constant as one walks towards it.

Perceptual constancy has been a central topic within psychology and philosophy for some time, but the term was coined by the Gestalt psychologists in the 1920s. Since different theorists use the term to refer to a variety of distinct yet related phenomena, it is important to be clear about how I am using it here. By 'perceptual constancy' I mean the phenomenon of some thing's properties (shape, size, colour, etc.) appearing to remain constant through differences in the way that the object seems with respect to that property. Understood in this way, perceptual constancy concerns appearances and is, therefore, something that clearly falls within the province of phenomenology. This phenomenological understanding of constancy is our topic (it contrasts with the related, but distinct, understanding of constancy as the discrepancy between a thing's apparent properties and the properties of the retinal image of that thing).

The phenomenological understanding of perceptual constancy poses a challenge for the description of things as they appear. How can it be that, for example, a thing's colour can be given as both constant and varied? Before looking at Husserl's and Merleau-Ponty's answers to this question, however, we should briefly consider the view that no such answer is required since there is in fact no such phenomenon as perceptual constancy, so described.

1.2 Scepticism about perceptual constancy

Bertrand Russell begins his well-known introduction to philosophy with the following description of the visual experience of a partly shaded table:

> Although I believe that the table is 'really' of the same colour all over, the
> parts that reflect the light look much brighter than the other parts [...] there
> is no colour which pre-eminently appears to be *the* colour of the table.
>
> (1912, p. 2)

Here Russell distinguishes between the way that the table appears and the way
that it is judged to be. He rightly points out that we judge the table to be the
same colour all over. However, he claims that this judgement cannot simply
result from one's taking the content of one's visual experience at face value
since, as he puts it, no single colour 'appears to be *the* colour of the table'.
With this, Russell is effectively denying the reality of colour constancy, at least as
understood phenomenologically. For colour constancy involves the appearance
as of a constant colour despite the differences to which Russell points.

This is a phenomenological disagreement between those who accept that
things, despite being variously shaded, sometimes really do *appear* to be the same
colour all over and those who, like Russell, deny this. We can formulate a similar
sceptical position with respect to the other properties for which perceptual constancy is often thought to apply. For example, it is often said that, as I move
around the table, it constantly appears rectangular in shape, despite the fact
that my experience of its shape is continuously changing. The constancy-sceptic
would deny this, asserting that, in such a situation, there is no single shape that
the table appears to be. Similarly, those who accept the phenomenological
conception of size constancy will say that, as I walk away from the table, it
appears to remain the same size. Again, the constancy-sceptic will deny this.

How are we to decide between the view that the table *looks* to have a single
colour (or shape, or size) and the view that we merely *judge* this to be so?
First, notice that the two views will give different answers to the following
question posed by Merleau-Ponty, who asks: 'on what basis, then, do we
judge that a form [shape] or a size are the form and size *of the object*?' (1945,
p. 312). If we accept that there is such a thing as perceptual constancy, we
may give the straightforward answer that our judgement that the thing, in this
case a table, has a certain shape (or colour, or size) is grounded in the fact
that it looks to have that shape (or colour, or size). That is, we take our visual
experience at face value. The constancy-sceptic, who denies that there is one
shape (or colour, or size) that the table appears to have, cannot offer this
answer. What alternative answer can the sceptic provide? Merleau-Ponty
suggests the following possibility:

> for each object we are given sizes and forms that are always variable according
> to the perspective, and [...] we agree to consider as true the size that we
> obtain of the object at arm's length or the form that the object assumes when
> it is situated upon a plane that is parallel to the frontal plane. These are no
> more true than others.
>
> (1945, p. 312)

On such a picture, our move from the ever varying properties presented in experience to the constant properties judged to truly belong to a thing is entirely conventional. As Russell describes the situation for the case of colour:

> there is no reason for regarding some of these [merely apparent colours] as more really its colour than the others [...]. When, in ordinary life, we speak of *the* colour of the table, we only mean the sort of colour which it will seem to have to a normal spectator from an ordinary point of view under usual conditions of light.
>
> (1912, p. 2)

Such views claim that although perceptual experience presents a thing's properties as constantly in flux we ordinarily treat the thing as truly possessing constant properties. In Russell's account of colour, this will be the colour that it displays in daylight. On the view considered by Merleau-Ponty, it will be the size at arm's length, or the shape when parallel to the frontal plane, respectively.

Russell goes on to point out that on his account, 'the other colours which appear under other conditions have just as good a right to be considered real'; thus our choice of constant colour is entirely conventional. He concludes from this that we must 'deny that, in itself, the table has any one particular colour' (1912, pp. 2–3). Exactly similar remarks can be made for shape, size, and any other property the experience of which involves perspectival variation.

The details of conventional accounts of constancy judgements need not detain us here since Merleau-Ponty offers a fundamental criticism of any sceptical view. He complains that such an account 'takes for granted what was to be explained, namely a range of *determinate* sizes and forms among which it would suffice to choose one, which would become the real size or the real form' (1945, p. 313). To see what Merleau-Ponty has in mind here, consider that the constancy-sceptic claims that the properties that I, 'in ordinary life', judge to be the true properties of an experienced thing are some subset of its merely apparent properties. For example, I will judge the tree beyond my window to be that size that it displays at arm's length. As the above quotation from Merleau-Ponty suggests, for such an account to yield the desired result, it must be that the 'merely apparent' shapes, sizes, and colours are given as determinate qualities. This is for the reason that the properties that we go on to judge to truly belong to the thing are themselves determinate.

It is not obvious, however, that this condition is met. As Merleau-Ponty puts it: 'for a single object that is moving away or that is spinning, I do not have a series of "psychical images," increasingly small or increasingly distorted, among which I could make a conventional choice' (1945, p. 313). Consider the experience of the tree before me. We must ask what merely apparent size the tree has in my current experience of it. How do I go about answering that question? I might measure it with my hands or, even better, with a ruler. But where do I hold the ruler? With arms outstretched, I will measure the apparent size of the tree at sixteen inches. If I move the ruler much closer to my face, it comes in at a mere six inches. Which of these is the apparent size of the tree? There appears to be no obvious non-conventional way in which to answer this question. The consequence of this seems to be that the apparent size of the tree is not given as a determinate, measurable quantity and so the sceptic's claim that experience presents us with only apparent sizes, one of which is conventionally selected as the 'true' size of the thing, must be rejected.

Whilst this will strike some as a compelling phenomenologically based argument against the sceptic about size constancy, it may seem less obviously so against the sceptic about constancies of shape and colour. The equivalent, for the case of colour, of the ruler measurement test would presumably involve holding up variously coloured test cards in order to find a subjective match. For the case of shape, one would need to hold up test cards that are variously shaped. This, it might seem, gives us a non-conventional way of divining the determinately given apparent properties of the thing.

Things, however, are not so simple. First, for the case of shape, are we assuming that the merely apparent shapes under discussion are two-dimensional? It is not obvious. Such a question might seem odd to ask at this point in the discussion, where the notion of a merely apparent shape has already played a significant role. But the very fact that it is not obvious how to answer the question suggests that merely apparent shape – that respect in which the appearance of an object alters shape-wise as it spins or I move around it – is not given as a determinate, measurable quality.

Second, for the case of colour, the implicit assumption above is that it will not matter where the test card is held (at arm's length or otherwise). Only in this way is the consequence avoided that merely apparent colour can only be given determinacy *via* a conventional choice. But this assumption is false. In any ordinary situation, the distance from my eye at which I hold the test card will affect the distance of that card from the light source and this, in turn, will affect the way that it looks colour-wise. As such, a test card the colour of which, when held at arm's length, subjectively matches some seen object, is unlikely to so match when I move it closer to my face. Once more, this suggests that merely apparent colour – that respect in which the appearance of an object alters colour-wise as the lighting changes – is not given as a determinate, measurable quality.

Whether or not the constancy-sceptic can satisfactorily answer this determinacy objection, there is further reason to doubt the sceptical denial of perceptual constancy. This is that there is reason to suppose that constancy cannot be reduced to judgement. One tactic often employed to determine whether some phenomenon is genuinely perceptual or is rather a matter of judgement is to consider whether or not it is belief independent. So, if I wish to know whether the leaves on the tree before me really *look* green, or if I merely judge them to be so, I may consider a situation in which I do not believe them to be green, perhaps taking myself to be subject to some sort of visual illusion. In such a scenario, we will presumably say, the appearance of the tree will remain the same, with only my response altering. Further, it is overwhelmingly plausible to suppose that, in this case, the leaves will continue to appear green. Given this, we may conclude that, in the actual situation, I do not merely judge the leaves to be green, they really do look green.

Striking examples of belief independence are given by visual illusions, for example the well-known Müller-Lyer illusion (Fig. 4.1), in which the two lines do not look of equal length despite the fact that one knows that they are.

Applying this to the case of perceptual constancy, it might be argued that even if I were to believe that a table is in fact cleverly painted in order to look as though it were partly shaded (or that it actually gets smaller as I move away, or that it actually changes shape as I move around it), it still would not look as though this were the case. That is, it would still look to be of one uniform colour. On these grounds, it might be said, we can reject the sceptical view.

Arguably, however, such a response to the constancy-sceptic begs the question. For that the table looks to be of one uniform colour is exactly what the sceptic denies. As such, the claim that, in the imagined scenario in which I believe the table to be cleverly painted, it will still look to be of one uniform colour, cannot be used as a premise in an argument against scepticism about perceptual constancy.

We can make some progress here by considering the image in Figure 4.2, known as the checkerboard illusion. Here we see a partly shaded checkerboard. Consider squares *A* and *B*. It is overwhelmingly natural to describe *A* as a black square and *B* as a white square. Put another way, *A* looks to be the same colour as the other black squares and *B* looks to be the same colour as the other white squares. This is despite the fact that both the black and white squares are variously shaded.

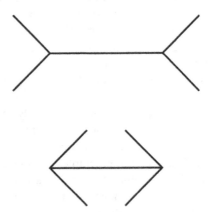

Figure 4.1 The Müller-Lyer illusion

Edward H. Adelson

Figure 4.2 The checkerboard illusion

Of course, the constancy-sceptic will reject this description, claiming that *B* looks darker than the unshaded 'white' squares. This might be supported by the fact that, in order to paint *B* one would need to use a different colour of paint than one would use if one were to paint one of the unshaded 'white' squares. But that thought can seem implausible when confronted with the fact that, in this sense that the constancy-sceptic is pushing, squares *A* and *B* are in fact the same colour, as can be seen from Figure 4.3.

If the fact that one would need to use different colours to paint the 'white' squares shows that they do not all look to be the same colour, then one would

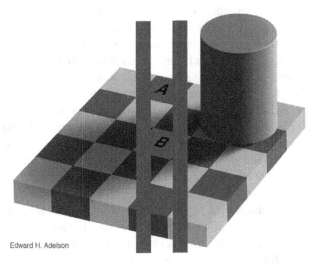

Edward H. Adelson

Figure 4.3 The checkerboard illusion revealed

expect that the fact that one would need to use the same colour to paint *A* and *B* would show that *A* and *B* look to be the same colour. In Figure 4.3 they do but it stretches credibility to suppose that the same is true of Figure 4.2. The addition of the two vertical lines in Figure 4.3 changes the way that squares *A* and *B* look. In particular it changes the way they look relative to each other. Since, in Figure 4.3, they look the same colour, it would seem to follow that they do not look the same colour in Figure 4.2. In which case, we have placed significant pressure on the constancy-sceptic's denial that the white squares, including *B*, all look the same colour.

Nothing said above shows that the sceptical position is false. The constancy-sceptic can consistently admit that squares *A* and *B* look different in Figure 4.2 whilst denying that, for example, *B* looks the same colour as the other white squares. We have said enough, however, to justify treating as default the view that there is such a phenomenon as perceptual constancy (for further discussion, see Schwitzgebel, 2006).

1.3 Constancy and memory

Merleau-Ponty considers an account of perceptual constancy in terms of memory, asking:

> what is the real colour and how do we have access to it? It will be tempting to respond that it is the colour according to which I most often see the table, the one that it takes on in daylight, at close proximity, under 'normal' conditions, in short, the most frequent conditions.

> (1945, p. 318)

Merleau-Ponty here bundles a number of non-equivalent claims together. Most obviously, the apparent colour that something displays in daylight need not coincide with that which it displays most frequently. Certainly not for those of us who spend too much time in artificially lit buildings. In any case, whilst an account in terms of frequency may work for some cases, it certainly would not for all. There are many types of objects, such as items of clothing, book covers, and so on, that have no typical colour and so no colour that they most frequently appear to have. Nor can the required frequency be a matter of how the *particular* thing in question usually looks, since we happily take previously unseen things to possess constant colours. Finally, and fundamentally, an account in terms of frequency – one that identifies the experienced constant colour of a thing with that 'which is predominant because it is inscribed in me by numerous experiences' (1945, p. 318) – is in fact subject to the determinacy objection that we have also seen is faced by the constancy-sceptic. For such a view must posit a learning period during which experienced things do not display constant colours. Only in this way will one merely apparent colour come to be the most frequently seen. But, once more, such a view must suppose that such merely apparent colours are given determinately, a supposition that we have seen some reason to doubt.

It is best, then, to focus on the idea in the above quotation that the colour that a thing appears to really have – its constant colour – is that which it displays in daylight. This is, of course, reminiscent of Russell's conventional account, indeed the two are not always clearly distinguished in Merleau-Ponty's discussion. The present view, however, is not a form of scepticism about perceptual constancy, but an account of it. The thought here is that a thing's constant colour is that colour, preserved in memory, that it displays in daylight. This is treated not as a conventional choice but as something given in perceptual experience itself. Once such a colour is installed within memory, it can then replace the merely apparent colour that the object may otherwise display when in shade, and so on. As Merleau-Ponty describes the view: 'I displace the actual color to the benefit of a color from memory [...]. The constancy of the color would thus be a real constancy' (1945, p. 318). Similar accounts might be described for both shape and size, accounting for shape and size constancy, as was the case for the conventionalist sceptic, in terms of the shapes and sizes that things present at arm's length, or when parallel to the frontal plane, respectively.

Merleau-Ponty's criticism of this view helps us to understand the motivation for the views of perceptual constancy, discussed in the following sections, that are offered by Husserl and by Merleau-Ponty himself. This is that even if the frequency view can account for perceptual constancy, it cannot account for the equally important phenomenon of perceptual variation; the feature of the experience of properties that helps to give rise to the puzzle in the first place.

The puzzle that perceptual constancy raises for the way things look was that, as viewing conditions change, a thing seems to vary yet remain constant with respect to one and the same property. So, as the afternoon light fades, the tree beyond my window changes colour-wise yet nevertheless appears to remain the same colour. However, as described above, the frequency account of perceptual constancy explains only the fact that it appears to remain the same colour: it appears that colour, preserved in memory, that it most frequently appears. But this view has lost sight of the fact that the tree appears to change colour-wise. As Merleau-Ponty reminds us:

> it cannot be said that the brown of the table is presented in all lighting conditions as the same brown, or as the same quality actually given by memory. A sheet of paper seen in the shadows, that we recognize as such, is not purely and simply white.

> (1945, p. 318)

A satisfactory account of perceptual constancy must accurately describe the way in which things' properties seemingly vary while remaining constant.

This is something that the memory account fails to do, explaining only the respect in which they remain the same.

2 Husserl on perceptual constancy

How can it be that a thing's properties can be given as both constant and varied? I sketched Husserl's answer to this in the previous chapter. As applied to the case of size, what I have been calling the sceptical position is familiar from the discussion there of the Humean argument for sense-data. Hume claimed that '[t]he table, which we see, seems to diminish, as we remove farther from it' (1739–40, p. 118). In the last chapter I outlined the Husserlian response to this view. On Husserl's view, the size of the table does not appear to change as we walk away from it. Rather, the table seems to remain the same and it is *how the table is adumbrated* that changes. That is, Husserl accepts the phenomenon of size constancy and accounts for the fact that one's experience of things nevertheless seems to change with respect to size by claiming that sizes are given through adumbrations. As Husserl writes: '[o]ne and the same shape (given "in person" *as* the same) appears continuously but always "in a different manner," always in different adumbrations of shape' (1913, p. 87).

2.1 Constancy and adumbration

The above is, however, only an initial sketch of an account and there is a good deal more to be said (cf. Madary, 2010). Recall Merleau-Ponty's question 'on what basis, then, do we judge that a form [shape] or a size are the form and size *of the object*?' (1945, p. 312). To see that we do not yet have an adequate answer to this question, consider again square *B* from Figure 4.2 above. Those who, like Husserl, accept that constancy is a genuinely perceptual phenomenon will agree that square *B* looks white. But they will also accept that it looks different colour-wise to the unshaded white squares. Applying Husserl's account of adumbration, we can say that whilst all of the white squares look the same colour, that colour is in each case given through a different adumbration of whiteness. It may seem, then, that we have answered Merleau-Ponty's question: we judge that the square is white on the basis of its' being presented as white through some adumbration or other. But whilst this goes some way towards answering the question, it is not a full answer. For, we will want to know, why is the square given as white, through a grey-ish adumbration, rather than as the grey colour that would in fact be used if one were to paint it? Or, to change the example to one concerning shape, why when I see an obliquely presented coin, does it look circular, with its circularity given through an ellipse-ish adumbration? Why does it not just look elliptical? Put another way, why is whiteness (or circularity), rather than greyness (elipticality), given as a property of the thing?

In response to this the Husserlian may claim that necessarily all and only intentional objects are adumbrated. When aware of an entity – including things, properties, and events – one is necessarily aware of it perspectivally and so through an adumbration. Thus, in the case of the obliquely presented coin, circularity is adumbrated, elipticality is not. This, it might be said, gives us an explanation of the fact that circularity is given as a property of the coin whilst elipticality is not. Such a fact, if it is one, is part of the reason for Husserl's claim that adumbrations are not intentional objects but are, rather, non-intentional, sensational features of experience.

It is, however, far from obvious that this point alone will enable the Husserlian to answer Merleau-Ponty's question in a satisfactory way, since it is not yet clear what grounds the claim that elipticality is not given *via* adumbrations. What substance can the Husserlian give to this claim?

2.2 Intuitive fulfilment, co-presence, inner horizon, and anticipation

It is at this point that a number of Husserlian notions, introduced in the previous chapter, come back into the picture. To see how, we first need to return to the experience of things which, of course, is intimately bound up with the experience of their properties that is our current concern.

Recall from the previous chapter that, since we only ever see a thing from one particular perspective, our view of it at any one time is always limited. Now, the facing side of a seen object is given with intuitive fullness. On Husserl's account this amounts to the claim that it is given *via* the animation of visual sensations. Nevertheless, those sides that I do not strictly speaking see are co-presented in visual experience. The co-presented is given *via* perceptual anticipations of further perspectival views of the thing, the perception of a thing as a thing necessarily involving this relationship between intuitive fullness and the co-given.

As we have seen, Husserl claims that a perceived thing's properties are also given through adumbrations. Thus, the whiteness of a square, or the circularity of a coin, will be given through adumbrations of colour and shape respectively. So, the shape and size of a visually perceived coin is given from a particular perspective; the colour of a visually presented square is given under particular lighting conditions. That a coin is given as circular means that circularity determines the law-like way in which one's perceptual anticipations of the coin are structured: if it is tilted away from one, then its elliptical-ish adumbrations will become increasingly 'thinner'; tilted towards one, then its elliptical-ish adumbrations will become increasingly 'fatter' up to the limit of being circular-ish. In line with Husserl's account of intentionality more generally, these adumbrations are 'interpreted' as appearances of circularity. The coin is thereby given as circular.

Finally, then, we can return to Merleau-Ponty's question. What grounds our judgement that the square is white, or that the coin is circular, is that it is given as white, or circular, through this law-governed system of perceptual anticipations. The square is given as white, not as grey, because its adumbrated appearance is perceptually anticipated to alter in accordance with the laws governing the relations between whiteness and, most obviously, lighting conditions. The coin is given as circular, not as elliptical, because its adumbrated appearance is perceptually anticipated to alter in accordance with the laws governing the relations between circularity, orientation, distance, and so on. Thus, it might be argued, can the Husserlian account of adumbration – with its associated notions of intuitive fulfilment, inner horizon, and perceptual anticipation – offer a satisfactory account of the perceptual constancies.

2.3 Husserl and 'intellectualism'

Throughout *Phenomenology of Perception*, including in his discussion of perceptual constancy, Merleau-Ponty criticises what he calls 'intellectualism', of which Husserl's account of perceptual experience might be argued to represent a variety. The question arises, then, as to whether the Husserlian account is subject to Merleau-Ponty's critique. Merleau-Ponty takes the term 'intellectualism' to pick out a number of related views, the central uniting feature of which is the claim that judgement is an element of – in some cases replaces – perceptual experience. Towards the beginning of *Phenomenology of Perception*, Merleau-Ponty points out that '[j]udgment is often introduced as *what sensation is missing in order to make a perception possible*' (1945, p. 34). In this way the intellectualist account of perceptual experience runs counter to the distinction that we ordinarily draw between the two:

> Between sensing and judging, ordinary experience draws a very clear distinction. It understands judgment to be a position-taking; judgment aims at knowing something valid for me across all the moments of my life and valid for other existing or possible minds. It takes sensing, on the contrary, to be the giving of oneself over to the appearance without seeking to possess it or to know its truth. This distinction disappears in intellectualism because judgment is everywhere that pure sensation is not, which is to say that judgment is everywhere.
>
> (1945, pp. 35–6)

With this in mind, recall the Husserlian account of perceptual experience as involving perceptual sensations which are 'animated' so as to constitute the experience of some thing and its properties. As Husserl puts it:

> Data are animated by '*construings*' within the concrete unity of the perception and in the animation exercise the '*presentative function*,' or as united with the construings which animate them, they make up what we call '*appearings of*' colour, shape, and so forth.
>
> (1913, p. 88)

This picture, with its 'construings' (or 'interpretations') looks very much like the intellectualist account that Merleau-Ponty seeks to reject. For what are *construings* if not judgements? Whether or not it is fair to think of Husserl as introducing judgement into perceptual experience, there is no doubt that such construings perform just the function that Merleau-Ponty's intellectualist takes judgement to play, namely transforming otherwise non-intentional sensations into perceptions of things and their properties. A Husserlian construing is precisely '*what sensation is missing in order to make a perception possible*'. There is, then, some reason to suppose that Merleau-Ponty's critique of the intellectualist account of perceptual constancy will, if successful, be a concern for the Husserlian.

As we saw above, Husserl's account has it that a thing is given as possessing a constant shape because its adumbrated appearance is perceptually anticipated to alter in accordance with the laws governing the relations between shapes, orientation, distance, and so on. Analogous remarks apply to size and colour. In this vein, Merleau-Ponty offers the following summary of the intellectualist view as it applies to size:

> the true size of my fountain pen [...] is the invariant or the law of corresponding variations of the visual appearance and of its apparent distance [...]. If I hold my fountain pen close to my eyes such that it conceals almost the entire landscape, its real size remains quite modest, because this fountain pen that masks everything is also a fountain pen *seen up close*, and this condition – always noted in my perception – restores the appearance to its modest proportions.
>
> (1945, p. 314)

This, however, is to get things back to front. As Merleau-Ponty says:

> When I see the furniture of my room in front of me, the table with its form and size is not, for me, a law or a rule for the unfolding of phenomena, it is not an invariable relation; rather, because I see the table with its definite size and form, I presume for every change in distance and orientation a corresponding change of size or form, and not *vice versa*.
>
> (1945, p. 315)

At least two objections to the intellectualist account can be found in Merleau-Ponty's discussion, both of which we have seen already. For the case of colour, Merleau-Ponty claims that the intellectualist account suffers from the blindness to variation with which he charged the memory account. That is, the intellectualist cannot account for the fact that, whilst a thing's properties appear constant, the experience of them nevertheless alters. We are prevented, he claims, from treating the constancy of colours as an ideal constancy and from relating it to judgment. For a judgment that would distinguish the contribution of the lighting in the given appearance would only come to an end with an identification of the object's proper colour, and we have just seen that its colour does not remain identical.

(1945, p. 318)

It is not clear, however, that this objection hits its target, at least not if that target is Husserl. For Husserl does not replace perceptual experience with judgement. Rather, he takes construing (which plays the role that the intellectualist gives to judgement) to be one element of perceptual experience, the other being perceptual sensation. His intellectualism is, we might say, in this respect modest. It is precisely changing perceptual sensation that Husserl takes to account for the variation experienced alongside perceptual constancy. Merleau-Ponty's objection, it seems, would perhaps have more bite against those who seek to do away with the notion of perceptual sensation altogether (that this is so is indicated by Merleau-Ponty's own rejection of 'pure sensation'. Note his remark, above, that for the intellectualist, 'judgment is everywhere that pure sensation is not, *which is to say that judgment is everywhere*', my emphasis). Whatever its other merits, Husserl's account of sensation provides him with a ready answer to this concern.

A second objection to the intellectualist rests on the claim that for perceptual constancy to be the result of the law-like relations between apparent shape, orientation, etc. is for those things to be given as determinate, measurable qualities:

> When it is said that true size or form are merely the constant law according to which appearance, distance, and orientation vary, it is implied that they could be treated as measurable sizes, and thus that they are already determinate, whereas the question is precisely to understand how they become determinate.
>
> (Merleau-Ponty, 1945, p. 315)

We have, of course, seen this objection before. We have seen that there is reason to doubt that merely apparent sizes, shapes, and colours are given as determinate, measurable qualities. If Merleau-Ponty is right on this score, and is also correct to suppose that the law-like connections employed by the intellectualist rely on such determinately given qualities, then the Husserlian account of perceptual constancy will turn out to be inadequate for by now familiar reasons.

3 Merleau-Ponty on perceptual constancy

What account does Merleau-Ponty himself provide of the experience of constant properties? From the discussion thus far we can draw the following constraints that Merleau-Ponty thinks any successful account must respect: it must be an account of the experience of constancy within variety, and it must not rely on the dubious assumption that merely apparent properties are given as determinate qualities. Although his account is difficult to pin down – and the interpretation I offer below is both partial and, in places, somewhat speculative – we can certainly distinguish a number of elements: the relation between the experience of a thing's constant properties and what we can call the experiential context; the role of normativity in the perceptual experience of a thing's properties; and the way in which one's experience refers back to one's bodily agency (for discussion of Merleau-Ponty's account, see Kelly 1999 and 2005a).

In what follows I follow Merleau-Ponty in first outlining the account of size and shape constancy before moving on to the constancy of colour. Since each of these places a significant emphasis on the experiential context of the perception of a thing and its properties, I begin with a brief discussion of this notion.

3.1 The experiential context

Husserl's account of constancy rightly emphasises the respect in which the way that a thing appears in perceptual experience is determined in part by the context in which it is situated. By 'context' here I mean not only what Husserl calls the external horizon – things in the surrounding environment – but also, and crucially, the spatial orientation (up, down, left, right, near, far) and lighting conditions of the perceptual field. One of the ways in which Husserl's intellectualist account goes wrong, on Merleau-Ponty's view, is in supposing that these latter are given as determinate magnitudes. Whilst this is, according to Merleau-Ponty, an error it would be just as serious a mistake to suppose that they are not given at all. Rather, they are given as elements of the ever-present background of all perceptual experience: the 'world-horizon', a notion clearly related to the Heideggerian account of the world that we are being-in:

> The constancy of colour [or shape, or size] is merely an abstract moment of the constancy of things, and the constancy of things is established upon the primordial consciousness of the world as the horizon of all our experiences.
>
> (1945, p. 326)

Merleau-Ponty claims, following the Gestalt psychologists, that all perceptual experience displays a figure-ground structure. When I experience an object it will be given against a background. So, as I look at the tree beyond my window, it is presented against the buildings that continue behind it and also my desk which is given in the extreme foreground. As I transfer my attention to the desk, the tree now becomes a part of the experienced background, and so on. In all perceptual experience, a thing is presented against such a context and this is significant for the proper account of perceptual constancy. The key claim is that this ever-present 'world horizon' partly determines both the constant properties that an object appears to possess, and the varying ways in which those properties appear. In the case of the perception of colour, this is done by the lighting conditions. In the case of shape and size, it is the work of spatial orientation itself.

3.2 The constancy of shape and size

The general shape of Merleau-Ponty's account is best seen from a distance and I begin, accordingly, way back at a feature of Heidegger's account of perceptual experience indicated previously. Whilst we have spent some time discussing the experience of properties, we have not asked what range of properties I am perceptually aware of things as having. Is perception limited to the experience of shape, size, and colour, as might be supposed from the discussion thus far, or does perception have a broader scope? Heidegger and, following him, Merleau-Ponty take the latter view. In *History of the Concept of Time* Heidegger distinguishes between the *environmental* and the *natural* thing. The latter is the thing as it might be characterised as it is 'in itself', independent of its context – for example, that a chair is wooden, solid, and so on. The former is the thing as it appears within a cultural setting – for example, that it was made in a factory. Heidegger claims that in perception we are primarily aware of the environmental characteristics of the thing:

> when I say in ordinary language and not upon reflection and theoretical study of the chair, 'the chair is hard', my aim is not to state the degree of resistance and density of this thing as material thing. I simply want to say 'the chair is uncomfortable.' Already here we can see that specific structures belonging to a natural thing and which as such can be regarded separately – hardness, weight – present themselves first of all in well-defined environmental characteristics.
>
> (1925, p. 38)

The claim here is that, in perception, we are primarily aware of things *as they figure in our culturally specific practical lives*. We experience the chair as uncomfortable, not just as possessing certain 'objective' characteristics such as shape and size. Indeed, such *meaningful* experience of things is the more fundamental phenomenon, with the experience of things as the bare possessors of shape, size, and so on being a derived phenomenon, abstracted from the rich awareness of environmental things. Later, in *Being and Time*, this becomes Heidegger's well-known claim that in everyday experience we are aware primarily of *equipment*, rather than *occurrent* entities. Recall from Chapter 2 the fundamental character, in his account of being-in-the-world, that Heidegger accords to the practical way in which we dwell in an environment. Equipment, or the environmental thing, is that which shows itself as something *in-order-to*, i.e. as that which is *for* something (Heidegger, 1927a, §15). A pen is equipment for writing, a fork is equipment for eating, the wind is equipment for sailing, and so on. Later still, something like this claim can be seen in the psychologist James Gibson's claim that we see things as possessing *affordances* (Gibson, 1979).

Of crucial importance for our purposes is how this general thought also applies to the experience of space. On Heidegger's picture, we experience space primarily in terms of the meanings that its constituent places have for us – 'The "above" is what is "on the ceiling," the "below" is what is "on the floor"' (1927a, p. 101) – and distances in terms of our everyday activities – 'We say that to go over there is a good walk, a stone's throw, as long as it takes to smoke a pipe' (1927a, p. 103). That is, spatial locations and distances are not given to us as measurable quantities but as estimations in peculiarly human terms.

This is a picture that Merleau-Ponty takes on, summarising this aspect of his position neatly with his insistence on 'a space which consists of different regions and has certain privileged directions; these are closely related to our distinctive bodily features and our situation as beings thrown into the world', and saying of experienced things more generally that '[c]lothed in human qualities, they too are a combination of mind and body' (1948, p. 56). Notice here a focus on the body. This is something that, despite Merleau-Ponty's charge of intellectualism, also figured in Husserl's account of adumbration and constancy (one turns a coin, after all, with one's fingers), but was left implicit in Heidegger's discussion. In fact, Merleau-Ponty places his account of the body and embodied agency – topics to be covered in more detail in Chapter 8 – at the centre of his account of the experience of space. On his picture, it is one's capacity for bodily activity that provides one with a spatial orientation. As he puts it:

> the body as agent [...] plays an essential role in establishing a level [*niveau*] [...]. What counts for the orientation of the spectacle is not my body, such as it in fact exists, as a thing in objective space, but rather my body as a system of possible actions, a virtual body whose phenomenal 'place' is defined by its task and by its situation.
>
> (1945, p. 260)

What Merleau-Ponty here refers to as a spatial 'level' is 'a certain possession of the world by my body, a certain *hold* my body has on the world', allowing me to 'inhabit the spectacle' (1945, p. 261). This level determines what counts as standing upright and what counts as sloping, what is here and what is there, these directions being given sense by the peculiarities of my capacity for bodily agency; seen in terms of how I am able to act within the perceived world (cf. Talero, 2005).

But the spatial level is not only a level in this 'spirit level' sense. It is also a *standard* or *norm*, as in when one does or does not reach some required level.

Thus, the spatial level will determine, for a thing of a given sort, what counts as its being the *right* way up and, correlatively, what will give me the *best* view of it, bringing it into 'maximum clarity' (1945, p. 261). That is, the establishment of a spatial orientation is partly responsible for the fact that in order to see the tree better, I should take a step back; in order to get the best view of an obliquely oriented coin, I should turn it to 'face' me. This is for the reason that, for any object, there is some privileged perspective from which it is best seen.

> For each object, just as for each painting in an art gallery, there is an optimal distance from which it asks to be seen – an orientation through which it presents more of itself – beneath or beyond which we merely have a confused perception due to excess or lack.
>
> (1945, pp. 315–16)

All of this gives Merleau-Ponty an account of what it is to see something as at some particular distance. 'The distance between me and the object is not a size that increases or decreases, but rather a tension that oscillates around a norm' (1945, p. 316). If spatial orientation is experienced as providing a level (norm) for the experience of things, and for each perceived thing there is a norm determining how it is best seen, then the experience of the distance to some thing is not the experience of a determinate, measurable quality but that of the extent to which the thing falls short of presenting the best view. That is,

> [v]ariations of appearance are not increases or decreases of size, nor real distortions; quite simply, sometimes its parts mix together and merge, sometimes they are clearly articulated against each other and reveal their riches.
>
> (1945, p. 316)

To see something as at great distance is for it, with its constant size, to appear too far away to be properly seen. Further, the claim that the spatial level is given in terms of one's 'body as a system of possible actions' gives us a way of understanding both apparent and constant shape and size. Thus, apparent shape and size are given in terms of what must be done if the perceived thing is to be brought to maximum clarity, i.e. is to present its best view. Thus, in the experience of a distant tree, it might be given as requiring a walk of 'as long as it takes to smoke a pipe' to reach. Only then will it 'reveal its riches'. Similarly, for a coin to be presented obliquely is for it to be presented as needing to be turned to 'face' me if it is to be best seen.

In the above examples, and quite generally, merely apparent shapes and sizes are not given as determinate, measurable qualities but only approximately, in terms of roughly specified bodily actions. The account thus satisfies the indeterminacy constraint set out above. Further, given that as I walk towards the tree, or turn the coin to face me, the actions required to bring it into maximum clarity change, the view also accounts for the fact of variation. But, of course, the constraint was to account for *constancy* in variation. How, on Merleau-Ponty's view, are we to account for the experience of constant properties?

In answer to this question, Merleau-Ponty gives a number of examples which, predictably, analyse the experience of constant properties in terms of the active bodily stance that one takes to them when they are given with maximum clarity. So:

> [t]he thing is large if my gaze cannot encompass it, small if it does so easily [...]. The object is circular if, when all of its sides are equally close to me, it does not require any change of the curve of the movement of my gaze, or if such changes can be attributed to the oblique presentation according to the knowledge of the world that is given to me with my body.
>
> (1945, p. 317)

Of course, as we have seen, a pen held too close to my face cannot be 'encompassed' by my gaze. But in such a case it will be given as falling short of the optimum viewing distance; 'its parts mix together and merge'. In such an experience, it will be given as, on the one hand, *needing to be moved away in order to be properly seen* and, on the other, as *being easily encompassed by my gaze once this is achieved*. As I move it away from my face, I experience the pen's size as a constancy within variation, since I experience the former as changing but the latter as remaining the same. Analogous remarks apply to the experience of shape. Merleau-Ponty's view, then, in incorporating our bodily engagement with the world into the experience of space and spatial characteristics (shape and size), offers us an account of shape and size constancy that satisfies both the constraints with which we started this section.

It is, however, certainly possible to question this account and it is worth briefly mentioning a few concerns before moving on to Merleau-Ponty's account of colour. Perhaps the most obvious worry concerns the plausibility of Merleau-Ponty's bodily analyses of both apparent and constant shape and size. The idea is that, for example, how a certain shape looks is a matter of its appearing to require certain actions in order to bring it into full view. But it might strike one as implausible that something seemingly as fundamental as the perception of shape in fact demands such a rich description. Not only does Merleau-Ponty saddle visual experience with content concerning (bodily) activity – as does Heidegger in his account of the environmental thing – he is also claiming that the content of visual experience refers to normatively (i.e. concerning how things *ought* to be) given further visual experiences (the *best* view). One might well wonder whether a more minimal account is possible.

Related to this is the concern that the account offered might seem phenomenologically inadequate in that there are differences between seeing something as needing to be turned to face one if it is to be seen properly and, as Husserl would put it, its looking elliptical-ish. There are, for example, fine-grained differences between the different elliptical-ish appearances that a coin presents as it is turned over, differences that seem difficult to capture with the rather blunt 'turned to face me'. More fundamentally, one might worry that Merleau-Ponty's bodily descriptions are simply not plausible in the first place. Whilst Merleau-Ponty is surely right that, for example, the visual experience of distance is not given in feet and inches, one might baulk at the idea that it is given in terms relating *only* to bodily action. There is, it might be argued, something objectionably reductionist about the view: the experience of the constant properties of worldly things is effectively reduced to an awareness of how I may engage with them. Of course, this worry is one that may also be raised to the Heideggerian (or Drefusian) account of being-in-the-world as a form of practical know-how, or absorbed coping, that is the most basic way in which we are directed toward things (see Chapter 2). These, then, are some of the challenges that Merleau-Ponty's account faces.

3.3 The constancy of colour

The key notion in Merleau-Ponty's account of colour constancy is that of lighting. We can get an initial sense of the significance of lighting when, looking at Figure 4.2, we note that the presence of a light source, whilst not explicitly represented in the picture, is crucial in determining how the white squares look. For whilst the white squares all look white, they also look different colour-wise and their doing so is intimately bound up with the implicitly presented lighting. How, then, is lighting given? Or, as Merleau-Ponty asks, 'What happens in the moment when a certain patch of light is taken as lighting, rather than as counting for itself?' (1945, p. 322). As an answer, he writes:

> The lighting directs my gaze and leads me to see the object, so in one sense it *knows* and *sees* the object [...] it seems to me that the spectacle is *visible in itself* or ready to be seen, and that the light that explores the foreground and the background, forms shadows, and thoroughly penetrates the spectacle accomplishes a sort of vision out in front of us [...]. We perceive according to light.
>
> (1945, p. 323)

The idea here seems to be that lighting is presented in vision as that which makes things visible, as a condition of the possibility of vision itself. Lighting is that which reveals things and their properties. Merleau-Ponty gives us an example to help us see this way in which lighting is given, contrasting it with the way in which a thing is. The light cast by a torch shining on a black disc is given as 'a whitish solid whose base is constituted by the disc' (1945, p. 321).

When, however, a white piece of paper is placed in front of the black disc, the beam of light is no longer given as solid – as a thing – but as that which makes the paper and disc visible with their respective colours. As Merleau-Ponty says, 'the beam of light and the disc are immediately dissociated and the lighting adopts the quality "lighting"' (1945, p. 325). When given as lighting, then, the beam of light is that which directs our gaze to the things seen. It is visually present, not in the way that things are but by way of forming a part of the context against which perceived things are visible.

How, though, does this implicitly given lighting affect the way in which a thing's colour is given? It is in his answer to this question that Merleau-Ponty reintroduces the normative dimension. Lighting is given not as a determinate, measurable quality, but rather as a norm that determines how a thing is *best seen*. Lighting, he tells us,

> is not on the side of the object, it is what we take up, what we adopt as a norm [...]. This is why it tends to become 'neutral' for us. The shadowy light in which we remain becomes so natural for us that it is no longer even perceived as shadowy.
>
> (1945, p. 324)

Considering the example of 'yellow' electric lighting, he writes:

> the yellow light, by taking on the function of lighting, tends to situate itself as prior to every color [...] objects distribute the colors of the spectrum according to the degree and to the mode of their resistance to this new atmosphere. Every color-*quale* is thus mediated by a color-function and is determined in relation to a level that is variable. The level is established, and along with it all of the colour values that depend upon it, when we begin to live within the dominant atmosphere.
>
> (1945, p. 324)

The colour that a thing is seen to have is determined, in part, by the 'dominant atmosphere' (level). Once we have become acclimatised to a certain lighting – once we 'live within' it – we will be presented with an array of objects of various colours and with a certain distribution of reflections and shadows. Consider again Figure 4.2. Some white squares are in shadow. This, on Merleau-Ponty's picture, amounts to their being lit in a way that falls short of the norm. An analogous point holds for the experience of reflections. That is, if lighting is experienced as providing a level (norm) for the experience of things and, as we saw above, for each perceived thing there is a norm determining how it is best seen, then the way in which some particular thing is lit – in shadow, exhibiting glare, etc. – is not the experience of a determinate, measurable quality but that of the extent to which the thing falls short of presenting its best view.

In line with his account of the spatial level, Merleau-Ponty claims that to 'live within' an atmosphere is to adopt a certain active bodily stance. For something to be given as inadequately lit is for it to call for, or solicit, a certain bodily response. Its appearance demands, for example, that, to see it properly, one must move towards it or tilt one's head in order to get a *proper* look. Thus merely apparent colours are accounted for in much the same fashion as are merely apparent shapes.

As for how constant colours are given, unfortunately Merleau-Ponty does not furnish us with the sort of examples that he does in the case of shape and size. He tells us that

> I say that my pen is black and I see it as black in the sunshine. But this black is much less the sensible quality of blackness than it is a dark power that shines from the object, even when it is covered over by reflections, and this black as a dark power is only visible in the sense in which a moral blackness is visible. The real colour remains beneath the appearances just as the background continues beneath the figure, that is, not as a quality that is seen or conceived, but rather as a non-sensorial presence.
>
> (1945, p. 319)

Merleau-Ponty's attempt here to account for colour constancy, and thereby satisfy the constraint of constancy within variation, is hardly illuminating. He gives us no clear indication of *how* constant colours are given, simply claiming that they are. Given his general picture, one would expect an account in terms of the active bodily stance that one takes to them when they are given with maximum clarity, but he doesn't seem to offer one. Of course, this is not to say that such an account cannot be given. One option is to say, as contemporary enactivists do, that to perceive a constant colour is to have (implicit) knowledge of the ways in which the merely apparent colours will change with lighting conditions and movement (Noë, 2004, Ch. 4). This, however, with its emphasis on knowledge, runs close to the intellectualist account that Merleau-Ponty is keen to avoid. Despite the richness and insight of his account of lighting, then, it seems that Merleau-Ponty owes us an account, in bodily terms, of the way in which constant colours are given through variation. As a result, his account of colour constancy is incomplete, to be filled in by those sympathetic to his view.

4 Conclusion

The experience of things is closely bound up with the experience of their properties. After all, there is no experience of a thing that is not an experience of it as being some way. But the experience of properties brings with it a phenomenological puzzle, the puzzle of perceptual constancy. Both Husserl and Merleau-Ponty claim that scepticism about perceptual constancy is not phenomenologically accurate. But Merleau-Ponty goes on to charge the Husserlian 'adumbration view' with the sin of intellectualism. Merleau-Ponty's own account, taking on a broadly Heideggerian picture of the primacy of practical directedness towards environmental things, locates a normative element in perceptual experience. Whether his account of perceptual constancy, and of the experience of properties more generally, is acceptable remains to be seen.

5   Experiencing events

as soon as we attempt to give an account of time-consciousness [...] we get
entangled in the most peculiar difficulties, contradictions, and confusions.

Husserl, *On the Phenomenology of the Consciousness of
Internal Time (1893–1917)*

Looking at the clock on my wall, I can see that the second hand is moving. If
I look for long enough, I will also see that the minute hand has moved.
Longer still and I will see that the hour hand has moved also. But the way in
which I come to know of the second hand's movement seems very different
from the way in which I come to know of the hour (and perhaps also the
minute) hand's movement. Not only can I see *that* the second hand has
moved, I can see its movement. The movement of the second hand is a central
aspect of the visual experience. This is not so with the hour hand. Although I
can see that it has moved, over a brief enough period of time my visual
experience would not noticeably differ if it were stationary. I come to know
that the second hand is moving because I can see it moving. I come to know
that the hour hand is moving by noting that its position has altered over the
period of time that I have been looking.

1 Events in experience

One way to describe the difference between the way that the movement of
second and hour hands relates to experience is to say that, in experiencing the
movement of the second hand, I experience not only the clock and its properties
(position of hands, etc.), but also the *event* of the second hand's movement. In
the previous two chapters we have considered the experience of things and of
their properties. Whilst these represent two distinct ontological categories they
are similar in the way in which they occupy time. Both things and their
properties persist through time. Events, on the other hand, do not persist
through time, rather they happen over time. One way to put this distinction is
to say that whilst things (although arguably not properties) may have *spatial*
parts, only events have *temporal* parts or, as I shall call them, phases. Only
events are temporally extended. Thus, to say the movement of the second

hand is an event is, amongst other things, to say that it happens over time. To say that the clock itself is a thing is to say that it does not happen over time, rather it persists through time. Our topic for this chapter, then, is the experience of events, conceived as things that *happen*.

Given the intimate connection between the notion of an event and that of the passing of time, this topic is often referred to as that of time-consciousness. As the mundane nature of the above example suggests, the experience of events is something with which we are all acquainted. Indeed, we have already appealed to it in the preceding two chapters. For despite the fact that things and their properties do not *happen*, the accounts of the experience of them that we have already discussed describe temporally extended stretches of experience in which they are presented. This is most explicit in the discussion of the Husserlian notion of perceptual anticipation, a concept that will reappear in this chapter. It seems, then, that the experience of events – time-consciousness – is something with which we are familiar.

I.I A puzzle about the experience of events

Despite this, in trying to think about the experience of events it can quickly come to seem deeply puzzling. We can get a sense of the puzzle of time-consciousness by considering the following intuitively plausible claims, each supported by an initial consideration of the phenomenology of the experience of events:

Events take time to happen, and are experienced as such (cf. Dainton's (2000, p. 129) 'phenomenal binding principle'). As suggested above, this is a feature that distinguishes events from entities, such as things and properties, that persist through time. In the experience of an event we are aware of this feature; aware of its constituent phases as spread out over time. Further, the experience of an event is unified in the sense that we are not simply aware of different phases, but are aware of them as flowing one into the next.

The experience of events is a perceptual phenomenon (cf. Dainton's (2000, p. 115) 'phenomenological constraint'). As we saw in the above example, the movement of the second hand seems to be something given in perceptual experience. In that example the experience is visual but the experience of events is also possible *via* the other senses. Consider, for example, the auditory experience of a melody being played.

The objects of perceptual experience are concurrent with the experience itself (cf. Miller's (1984, p. 107) 'Principle of Presentational Concurrence'). At any point in time, a person is aware of what is happening right now, at the moment of the experience itself. Of course, depending on the distance from the perceived object, it takes a period of time for light- or sound-waves to reach the perceiver. Ignoring that complication, however – which in any case is not a concern from within the phenomenological reduction – we can say that perceptual experience is *and seems to be* concurrent with that which is perceived. In this respect, perceptual experience differs from memory. If I recall this morning's breakfast, the things, properties, and events that I bring to mind do not seem to occupy the same period of time as my memory-experience of them.

The puzzle, now, is that we seem to be perceptually aware of something both as being spread out over time and also as happening now. If, at any given time, we are aware only of what is happening now, how can we be aware, at any one time, of an event – something that, essentially, happens over time? If this seems impossible then the above line of thought must have made a mistake for, as testified by our opening example, there surely is such a thing as the experience of events. In what follows we shall look at a number of accounts of time-consciousness that, in one way or another, attempt to solve this puzzle.

1.2 Succession and unity

An initial reaction to the above puzzle may be that it is a spurious one generated by the fact that it is limited to what one can experience 'at any one time'. At any one time, it might be agreed, we are aware only of what is occurring at that time. But we have a constant stream of experience. Each momentary experience is immediately followed by another. Experiencing the continuous movement of the second hand, it may be said, is simply a matter of each momentary experience being an experience of the hand as occupying a slightly different position. Since a stretch of experience is constituted by a consecutive series of such momentary experiences, such a stretch will amount to an experience of the second hand as moving. An analogy with film makes this clear: whilst each frame contains only a static image, viewing frames one after the other at sufficient speed generates the experience of movement.

It is widely held that such a simple reply is not an adequate response to the puzzle. The reason for this is brought out forcefully by William James (1842–1910) in his hugely influential account of time-consciousness, who writes that a '*succession of feelings, in and of itself, is not a feeling of succession*' (James, 1890, Vol. I, p. 628). The point is elaborated by Husserl in a way that suggests where he thinks the solution lies:

> When a melody sounds, for example, the individual tone does not utterly disappear with the cessation of the stimulus or of the neural movement it excites. When the new tone is sounding, the preceding tone has not disappeared without leaving a trace. If it had, we would be quite incapable of noticing the relations among the successive tones; in each moment we would have a tone, or perhaps an empty pause in the interval between the sounding of two tones, but never the representation of a melody.
>
> (1905, p. 11)

The fact, that is, that the experience of an event consists of a series of experiences of the successive phases of that event, does nothing to explain the fact that the experience of an event is *unified*; that it is an awareness of those phases as flowing one into the next. For us to have an experience of an event *as an event*, it seems, there must be, at any one moment, some sort of awareness of the just past phases of the event simultaneous with the awareness of the present phase. This has been dubbed by Miller (1984, p. 109) the Principle of Simultaneous Awareness and is a central aspect of James's own 'specious present' view.

1.3 The specious present

James denies that the objects of an experience are all concurrent with the experience itself. We are, that is, experientially aware of a span of time that includes the present and immediate past (and perhaps future). This span is the 'specious present', an idea articulated by James in a well-known passage:

> the practically cognized present is no knife-edge, but a saddle-back, with a certain breadth of its own on which we sit perched, and from which we look in two directions into time. The unit of composition of our perception of time is a *duration*.
>
> (1890, Vol. I, p. 609)

On such a view the 'now' or 'practically cognised present' possesses duration, extending beyond the temporal location of the experience itself. One way to read this is to suppose that the sense in which the specious present is *present* is that the phases of the perceived event that are simultaneously experienced within the specious present are all experienced as happening *now*. Described in this way, however, the view seems to have the consequence that we experience the successive phases of an event as *simultaneous*. That is, the second hand would be simultaneously experienced as being at more than one place. This is for the reason that if two phases both *now* seem to happen *now*, then surely they both seem to happen at the same time. But this consequence is obviously unfaithful to our experience of time. It cannot be that those phases occupying the 'later' parts of the specious present are given to consciousness in just the way that those phases occupying the earlier parts are. As Husserl puts it:

> If, in the case of a succession of tones, the earlier tones were to be preserved just as they had been while at the same time new tones were to sound again and again, we would have a simultaneous sum of tones in our representation but not a succession of tones. There would be no difference between this case and the case in which all of these tones sounded at once.
>
> (1905, p. 13)

A closely related criticism is that, given that our experience at each moment would span a duration longer than a moment, we would presumably experience everything more than once. In a sequence of notes *c, d, e* we would experience *c* at the time at which *c* occurs, and then again at the time at which *d* occurs. But, of course, experience is not like this at all. We only experience each note once.

Whether these are fair criticisms of James, of course, depends on whether we have interpreted his view accurately and, fortunately for him, there is reason to suppose that we have not. Speaking of the specious present, James tells us that

> [i]ts content is in a constant flux, events dawning into its forward end as fast as they fade out of its rearward one, and each of them changing its time-coefficient from 'not yet,' or 'not quite yet,' to 'just gone,' or 'gone,' as it passes by.
>
> (1890, Vol. I, p. 630)

Thus, James does not suppose that phases simultaneously experienced within the specious present are experienced as *happening now* and so as simultaneous. Rather, as a phase 'moves' through the specious present the way that it is given changes correspondingly, from *not yet* to *gone* and all that lies between. Of course, this does leave it somewhat mysterious why the specious present should be thought of as the specious *present*, but I will leave that worry to one side.

If it is the case that those phases of an event that are, at any one moment, experienced within the specious present, are not experienced as happening *now*, what distinguishes them from phases that are so far away from the present moment so as to no longer be within that window? If I watch the second hand move for thirty seconds – say, from XII to VI – when it reaches VI we can assume that the experience of its position at V is still within the specious present. But what distinguishes this from my memory of the experience of its position at III? The answer, as will be anticipated by the initial puzzle of time-consciousness, is that the former, but not the latter, is a perceptual experience. As James puts it:

> the reproduction of an event, *after* it has once completely dropped out of the rearward end of the specious present, is an entirely different psychic fact from its direct perception in the specious present as a thing immediately past.
>
> (1890, Vol. I, p. 630)

James's account, then, of the specious present has it that, at any one moment, I am *perceptually* aware of moments both to come and just past. It is within this window of the specious present that we find the perceptual experience of events. This, however, is apt to seem mysterious (Kelly, 2005b, p. 230). How can it be, one may well wonder, that one has a direct perceptual access to the past (remember that we have put aside issues raised by the time-lag caused by the perceptual process itself)? This mystery is only compounded when we turn to the 'forward end' of the specious present. Surely we are not to believe that each of us can literally see into the, admittedly very near, future? One way to sharpen these concerns is to maintain that, unlike the present, neither the past nor the future phases exist, at least not at the time of the experience itself, and that James's view therefore entails that we can perceive non-existent things. This, it might be thought, is highly dubious.

A second worry about James's account is that it entails that our experience systematically misleads us as to its own temporal relation to experienced events (Phillips, 2010). For James's view rejects the claim, with which we began, that the objects of perceptual experience are concurrent with the experience itself. According to James we are, at a moment, aware of more than the phase of the event that is happening at that moment. Thus, the temporal location of the experience is not the same as the temporal location of that which is experienced. It follows that there is a systematic mismatch between the time at which an experience occurs and the stretch of time experienced.

Although this is a counterintuitive result, and one that the Jamesian will presumably find unwelcome, it is tempting to suppose that for our purposes it may be ignored. After all, we are attempting to provide a phenomenological account of time-consciousness and, at least as far as Husserlian phenomenology is concerned, that should be undertaken within the confines of the phenomenological reduction. As such, the location of both events and experiences in objective time must be placed in brackets and neglected. The phenomenologist, rather, is interested in 'appearing time, appearing duration, as appearing' (Husserl, 1905, p. 5). As Husserl points out:

> Someone may find it of interest to determine the objective time of an experience [...]. It might also make an interesting investigation to ascertain how the time that is posited as objective in an episode of time-consciousness is related to actual objective time [...]. But these are not tasks for phenomenology.
>
> (1905, p. 4)

Tempting as it may be this response is, in fact, unsatisfactory. There is an objection here that the phenomenologist must take seriously. For the claim that the objects of perceptual experience are concurrent with the experience itself was not just pulled from nowhere but was supposed to be phenomenologically grounded – based on how our experience seems to us. This means that the above objection can be rephrased as showing that rather than accurately accounting for the phenomenology of the experience of events, James's view is at odds with a central element of that phenomenology. That is, we started with the claim that our experience seems to be concurrent with that which is experienced. Now, construed as an account of the phenomenology, James's view has the consequence that experience and events are not concurrent in this way. It plausibly is a consequence of this that our experience does not *seem* to be concurrent with events experienced. If this is a consequence then James's account gets the phenomenology wrong. What we would need from the account would be an explanation of how apparent concurrence is consistent with the specious present or, to put it more generally, with the Principle of Simultaneous Awareness. This is something that we have not yet been offered.

2 Husserl on time-consciousness

Husserl never reached a settled view on the nature of time-consciousness, trying out different accounts at various points. In what follows I focus exclusively on his 1905 'Lectures on the Phenomenology of the Consciousness of Internal Time'. Husserl's position in these lectures is not altogether unlike James's. Like James, Husserl maintains that one's perceptual experience at any one moment is of a period of time that includes but is not limited to that moment. In this respect, Husserl's position can be thought of as a version of the specious present account (Dainton, 2010, but see Kelly, 2005b for an opposing interpretation). One advantage that Husserl's account may have over James's, however, is its ability to deal with the concerns raised at the end of the previous section. In what follows, I outline and then evaluate Husserl's view. As will become clear, even if Husserl can avoid the objections to which the Jamesian position is open, there are other serious concerns about his account.

2.1 Primal impression, retention, protention

Husserl claims that, in the experience of an event, one has at any one moment an experience not only of the phase occurring at that moment, but also of the phases that have just occurred, and the phases that are soon to occur. His labels for these three aspects of experience are, respectively, 'primal impression', 'retention', and 'protention'. All three must be in place for one's experience to present something as an event.

The primal impression is an intentional awareness of the present phase of an event as *now happening*. Taking the example of an experienced tone, Husserl describes it as 'the "source-point" with which the "production" of the enduring object begins', pointing out that it is 'in a state of constant change: the tone-now present "in person" continuously changes into something that has been' (Husserl, 1905, pp. 30–1). Relating primal impressions to retention, Husserl tells us that

> [d]uring the time that a motion is being perceived, a grasping-as-now takes place moment by moment [...]. But this now-apprehension is, as it were, the head attached to the comet's tail of retentions relating to the earlier now-points of the motion.

> (1905, p. 32)

Each such retention (or 'primary memory') is an intentional awareness of a past phase of the event as *having just happened*, as 'something that has been'. As such, each retention was once a primal impression: '[t]he tone-now changes into a tone-having-been; the *impressional* consciousness, constantly flowing, passes over into ever new *retentional* consciousness' (1905, p. 31). Retention 'holds in consciousness what has been produced and stamps on it the character of the "just past"' (1905, p. 38).

Husserl spends much less time on protention (or 'expectational intuition'): the intentional awareness of the future phase of an experienced event as *about to happen*. It is clear enough, however, what he means by the term. He writes: 'every process that constitutes its object originally is animated by protentions that emptily constitute what is coming as coming, that catch it and bring it toward fulfilment' (1905, p. 54).

In the unified experience of an event as an event, then, we can at any one moment discern three elements of the overall intentional awareness: we continuously experience a phase of the event as *happening now*, the phase having previously been emptily protained as *on its way*, and continuously 'moving' into the past and retained as *past*. Thus, at any one moment, we are aware of a duration of time that includes but extends beyond the time at which that experience is located.

So far I have focused on the way in which the phases of an event are experienced. But in the experience of an event, not only is the past phase of the event retained, so is the past experience – the primal impression – of that past phase. As I watch the second hand move from one position to another, not only do I have a retentional awareness of its earlier position, I also have a retentional awareness of my experience of its being at that position.

> The perceptual act sinks backward in time just as what is perceived in its appearance does, and in reflection identically the same temporal position must be given to each phase of the perception as is given to what is perceived.
>
> (1905, pp. 74–5)

This structural feature of time-consciousness – the awareness of one's immediately past experience – constitutes, then, a primitive variety of self-awareness, the topic of Chapter 7.

2.2 The temporal horizon

Despite discussing it only briefly, the notion of protention actually plays a fundamental role in Husserl's phenomenology as a whole. In fact, we have already encountered it in Chapters 3 and 4 in the guise of perceptual anticipation. Recall that, on Husserl's account of the perception of things, the fact that perceptual experience is perspectival means that not every aspect of an object is present to vision in the same way. In particular, some parts of a perceived object, although being visually present, lack 'intuitive fullness'. Such parts are given – as Husserl says 'co-given' – *via* perceptual anticipations in which I anticipate how the thing would look if, for example, I move towards or around it, or if the thing itself moves. Those parts of the perceived thing that are merely co-given form an 'inner horizon' of the experience.

The notion of an experiential horizon unites the experience of things with our current topic of time-consciousness. For such perceptual anticipations of how a thing will appear just are protentions. Put in this way we can see how Husserl's account of time-consciousness relates to his account of perceptual experience more generally. We see that Husserl's account can be described by saying that in any experience of an event, the phase presented as happening now is experienced against a horizon of those phases yet to happen and those just past. More crucially, we see that the analysis of time-consciousness forms the bedrock of his account of the perception of things and their properties. For it is by way of perceptual anticipation, and hence time-consciousness, that the experience of an entity as a thing or as a property is so much as possible. The experience of events enables the experience of an objective world. Without this experience of the continuous flowing of one phase into another, we would be aware of nothing but a jumble of sensations. It is no surprise, then, that Husserl regarded the analysis of the experience of time as the most fundamental aspect of the entire phenomenological enterprise.

This relation between retention and protention, on the one hand, and the general notion of a perceptual horizon, on the other, makes clear the sense in which Husserl treats the experience of events as perceptual. Just as, in the experience of things, only one aspect – the facing side – is given with intuitive fullness, so in the case of the experience of an event, only one phase – the present phase – is given in this way. Those phases of an event that are retained or pro-tained are, like the far sides of seen things, merely co-given. On Husserl's picture, as we saw in Chapter 3, this is cashed out in terms of perceptual sensations. So, in the temporal case, only the present phase – given through the primal impression – is presented by way of perceptual sensations. That which is co-given, for example in retention, is not 'sensed' in this way. If it were, it would appear as present rather than past, as is an echo or reverberation. As Husserl says:

> retentional 'contents' are not at all contents in the original sense. When a tone dies away, it itself is sensed at first with particular fullness (intensity); and then there follows a rapid weakening in intensity. The tone is still there, still sensed, but in mere reverberation. This genuine tone-sensation must be distinguished from the tonal moment in retention. The retentional tone is not a present tone but precisely a tone 'primarily remembered' in the now: it is not really on hand in the retentional consciousness.

(1905, p. 33)

So, just as in the case of the experience of things, Husserl's account of the experience of events relies on a distinction between two different ways in which things are given in experience: the present phase is given with intuitional fullness, the past (and future) phases are co-given with intuitional emptiness. Together, they amount to the perceptual experience of an event as something spread out over time.

Matters are complicated slightly by the fact that Husserl (1905, §§16–17) recognises two distinct uses of the term 'perception'. One treats all of primal impression, retention, and protention as perceptual, distinguishing them from the non-perceptual recollection of an event. Another treats only the primary impression as perceptual, distinguishing that from retention and protention. In the above discussion, I have employed the term in the first sense, the sense in which '[p]erception [...] is the act that places something before our eyes as the thing itself, the act that *originally constitutes* the object' (1905, p. 43). This is Husserl's way of accounting for the fact that, to return to our original example, the movement of the second hand seems to be something given in perceptual experience (cf. Dainton's (2000, p. 115) 'phenomenological constraint').

2.3 Retention and memory

Husserl refers to retention as primary memory, distinguishing it from recollection, or 'secondary memory'. Since views that attempt to explain time-consciousness in terms of memory are often dismissed rather swiftly (e.g. Dainton, 2000, Ch. 5), it is important to see what this distinction amounts to. In what sense, then, is retention distinct from recollection, conceived of as a 'reproductive' form of memory? Husserl is quite clear that there is an important distinction: 'the modification of consciousness that converts an original now into a *reproduced* now is something entirely different from the modification that converts the now, whether original or reproduced, into the *past*' (1905, p. 48).

There are, according to Husserl, a number of ways to draw this distinction. Perhaps the most obvious concerns the 'freedom' of recollection (1905, §20). That is, whilst the movement from primal impression to recollection is always automatic, the recollection of a past event is something that can be achieved at will. In recollection, we can 'run through' a past stream of experience quickly or slowly, and as often as we wish. None of this applies to retention. A second difference concerns the obscurity with which recollection presents its objects (1905, §21). Whilst a perceived event can be presented obscurely, perhaps seen through mist, there is a distinct sense of obscurity possessed by recollection which is not a matter of the object, but rather they way in which it is given. All recollected events, Husserl suggests, lack the clarity of perceptual experience (including retention) and are 'as if seen through a veil'.

These differences, one might object, are merely contingent. One can easily conceive of a perfectly clear memory that lacks the 'freedom' with which our ordinary recollections are imbued. There is, however, a more fundamental, structural difference between retention and recollection. Retention is essentially a part of a larger whole of which primal impression and protention are the other elements. In particular, retention is only conceivable as following on from a primal impression. Since 'every retention intrinsically refers back to an impression', it follows, according to Husserl, 'with *a priori* necessity that a corresponding perception, or a corresponding primal impression, precede the retention' (1905, p. 35). In fact, as Husserl describes it, the notions of retention and primal impression can only be understood in terms of each other:

> the now-phase is conceivable only as the limit of a continuity of retentions, just as every retentional phase is itself conceivable only as a point belonging to such a continuum; and this is true of every now of time-consciousness.
>
> (1905, p. 35)

Since experience is continuously flowing, the notion of a *moment* at which we can isolate primal impression, retention, and protention, is an abstraction. In fact, we are presented with a continuous renewal of the now, with events 'sinking' further and further into the past. Any particular retention is conceivable only as an element in such a continuous 'movement'.

Nothing like this can be said of recollection in which, by contrast, I reproduce a stretch of experience that contains the full protention-primal impression-retention structure. Speaking of recollection, Husserl tells us that

> [t]he whole process is a re-presentational modification of the perceptual process with all the latter's phases and stages right down to and including the retentions: but everything has the index of reproductive modification.
>
> (1905, p. 39)

So whilst retention is essentially an aspect of a present perceptual experience, contributing to the experience's amounting to an experience of an event as there 'in person', recollection is no such thing. Rather, recollection is the bringing to consciousness of an entire run of experience that is experienced as being in the past. Although, in recollection, we do recall an event from a particular temporal perspective, with a certain privileged point being represented as now, this is a past-now.

In fact, when I recollect an event, that act of recollection itself forms a part of my present experience and, correspondingly, generates a retentional awareness of the memorial consciousness as it runs through the different phases of the event. Recollection is, in this respect, a significantly more complex phenomenon. So, if I experience an event consisting in the phases A, B, C, I will at the time at which C occurs, retain both A and B. In fact, I will also retain the movement from A to B, that is I retain B *as following on from A*, a situation that Husserl writes as '$A-B$'. Thus, the stretch of my experience can be represented as follows

A

B (A)

C (A–B)

Here, what is included in the brackets is that which is retained. Thus, as *B* is given through primary impression, I retain *A*, as *C* is given, I retain *A–B*. If, however, I later run through the experience in recollection, I will remember each stage of the experience, complete with retentions, and also, since the recollection is itself an experience, I will retain the recollection itself. Thus, following Husserl in using *A'* to represent the recollection of *A*, and using square brackets to represent a complex but unified phenomenon – so, for example, *[B (A)]'* represents the recollection of having experienced *B* whilst retaining *A* – my experience can be represented as follows:

A'

[B (A)]' (A')

[C (A–B)]' (A'–B (A)')

Represented in this way, we can see that on the Husserlian account, the structure of recollection is quite different from that of retention. Retention is the more basic phenomenon, itself built into the structure of recollection.

3 Is Husserl's account tenable?

Whilst Husserl's account of time-consciousness contains a great deal more by way of subtle detail, the discussion of the previous section outlines its basic elements. The Husserlian account has, of course, been subject to a variety of objections (an influential example of which is Derrida, 1967, Ch. 5). Although I will not be able to address them all, in the present section we will consider the extent to which the account is able to withstand a number of these. In the process, in particular in response to the concurrence objection, we will see how Husserl's account can be interpreted as blurring the distinction between what Dainton (2010) has called the retentional and extensional models.

One criticism of Husserl's account of time-consciousness that I will not address in detail nevertheless deserves explicit mention here. This is Heidegger's suggestion that Husserl's is a 'vulgar' conception of time that misses the more fundamental respect in which we 'dwell' temporally. This conception, according to which 'time shows itself for the vulgar understanding as a succession of constantly "present" nows that pass away and arrive at the same time', is lacking for the reason that 'both datability and significance are lacking [...]. The characterization of time as pure sequence does *not* let these two structures "appear". The vulgar conception of time *covers* them *over*' (Heidegger, 1927a, p. 401).

In short, the Husserlian account operates with a levelled off conception of time, with no essential role for the fact that 'now' is *today*, or *Wednesday*, or *breakfast time*, and so on; no role for the fact that durations are measured out as, for example, 'as long as it takes to smoke a pipe'. In seeking to account for the experience of the passing of time in such a levelled off manner, Husserl is wrongly assuming that *this* is the most fundamental way in which time is given to us. But this, argues Heidegger, is a mistake. In much the same way that, on Heidegger's account, worldly objects are most fundamentally given in terms of their meaningful relations to one's own goals and activities, so time is given in terms of such significance – in terms of what it means to us. Heidegger's conception of time *in its significant relations to us* is, he claims, the fundamental one from which the levelled off Husserlian one is but an abstraction 'cut off from these relations' (1927a, p. 401). That is, just as equipment is the more basic phenomenon than occurrent things, so the time in which we dwell is more elemental a phenomenon than the series of 'nows' with which Husserl is concerned.

To do justice to this critique of the foundations of Husserl's account of time-consciousness would require us to investigate Heidegger's account of temporality in detail and evaluate the Heideggerian critique of Husserlian phenomenology more generally (see Blattner, 1999a, Ch 2–4; Keller, 1999). Unfortunately, the enormity of that task means that this is a critique that we must set to one side, instead focusing on a range of other objections.

3.1 Simultaneity

On one understanding of James's view, seemingly in accordance with the fact that his is an account of the specious *present*, every phase of an event that is experienced at a time is experienced as happening *now*. Given that the view also maintains that phases of the event that in fact occur at distinct times are all experienced at a moment, alongside the plausible thought that if two phases are now experienced as happening now then they are experienced as happening at the same time, there arises a problem. For it seems that such a view would conflate the experience of a sequence of distinct notes with that of a chord. Husserl is aware of this problem and explicitly distances his own account from such a picture, pointing out that the 'representation of succession comes about only if the earlier sensation does not persist unchanged in consciousness but is modified in an original manner' (1905, p. 13). His own account of such modification is, of course, the distinction between primary impression and retention. The latter, being an awareness of something *as past*, is not an awareness of it as happening now. Thus his account is not open to this simple objection.

3.2 Repetition

The same point provides an answer to the objection that specious present views entail the unwanted repetition of experienced events. Thus, if I am, at one moment, aware of a phase as occurring now and, at the next moment, still aware of it as occurring now, then it would seem that I would experience that phase as occurring twice. That is, I would be aware of the phase as happening at more than one moment. The fact that, on Husserl's account, I am not aware of a retained phase as happening now but as just having happened, answers this concern. It is true, on this view, that there is more than one time at which I experience the phase but it is not true that I experience the phase as occurring at more than one time. So there is no problematic repetition.

3.3 Non-existence

Recall the worry that James's view – according to which we are perceptually aware of the past and future phases of an event – mysteriously allows that we can perceive things that do not currently exist. In Kelly's words, 'it is hard to understand how I could now be perceptually aware of something that is no longer taking place' (2005b, p. 230). Whilst this objection may be a concern for some defenders of specious present views, it is not one that ought to trouble Husserl. For, recall from Chapter 3, Husserl's account of perceptual experience is an intentional one and intentional accounts of perceptual experience are, in principle at least, able to give an account of how it is that our awareness can present an object that does not (currently) exist. In such cases, for example those in which we are subject to hallucination, we have a merely intentional object before us. Of course, Husserl would not wish to place the retentions and protentions partly constitutive of the experience of an event in the category of hallucination. After all, it seems safe to assume that the content of a retention will typically (if not always, cf. Husserl, 1905, §13) be accurate – in most cases the phase in question *will* have just happened. The point, rather, is that Husserl's account of perceptual experience does not even require that the objects of experience exist, let alone that they exist at the time of the experience itself.

3.4 Concurrence

The concurrence objection, introduced during the discussion of James in §1.3, is more serious. This is grounded in the fact that reflection reveals that the experience of an event is concurrent with the event experienced. James's specious present view seemed to flout this, presenting an account according to which the experience of an extended duration of time in fact occurs at an instant. On the face of it, Husserl is open to the same objection. For, on his view, at any one moment, I am aware of phases that occur at other moments.

Furthermore, as we saw in §2.1 Husserl accepts (at least in his 1905 lectures) that experience is presented as concurrent with that which is experienced. Does this not leave his account of time-consciousness in a contradictory situation? Drawing on Phillips's (2010) presentation of this objection, we can say that, given concurrence, the perceptual experience of a duration of time will itself seem to possess a duration. But, on Husserl's account, in accordance with the Principle of Simultaneous Awareness, the perceptual experience of duration is available at a moment. Thus concurrence is not respected.

Husserl will, I think, respond to this objection by denying that, on his view, the perceptual experience of duration is in fact available at a moment. After all, as we have already seen in §2.3, Husserl claims that each retention (or, for that matter, primal impression) is conceivable only as a phase of a continuous experience. From this, Husserl draws the conclusion that every retention presupposes a preceding primal impression, claiming it as 'evident' that there is 'a law according to which primary memory is possible only in continuous annexation to a preceding sensation or perception' (1905, pp. 34–5). Again, he writes:

> when we have a retention of A [...] we by no means assert the having of the retention as evidence that A must have preceded it; but we do indeed assert it as evidence that A must have been perceived [...]. Within a transcendent perception, the immanent succession that belongs to its structure essentially is also absolutely certain.
>
> (1905, pp. 35–6)

The idea here seems to be that from the fact that the experience of succession is absolutely certain, and so survives the phenomenological reduction, we can conclude that there are in fact a multiplicity of moments, each possessing the protention-primal impression-retention structure. This, seemingly, is for the reason that primal impressions and retentions are merely abstractions from the continuous flow of experience. Whether or not Husserl is in fact justified in making this assertion, we can see that he is here denying that the experience of duration (or succession) is possible at a moment. Rather, such experience requires a stretch of experience over time. This time is not objective time, it should be remembered, but 'appearing time, appearing duration'. As Husserl writes:

> These are absolute data that it would be meaningless to doubt. To be sure, we do assume an existing time in this case, but the time we assume is the *immanent time* of the flow of consciousness, not the time of the experienced world.
>
> (1905, p. 5)

This is a significant point about Husserl's account of time-consciousness. It is often assumed that Husserl subscribes to what Phillips (2010, p. 180) calls the

Strong Principle of Simultaneous Awareness, according to which the unified experience of an event as happening over time can be had at a moment. The idea would be that a momentary experience, possessing the protention-primal-impression structure, is sufficient for the experience of a temporal flow. The above discussion suggests that this is not so. Rather, Husserl seemingly accepts what Phillips calls the Weak Principle of Simultaneous Awareness, according to which the unified experience of an event as happening over time is not possible at a moment, but requires a series of moments, each of which possesses a structure that is intentionally directed towards a period of time. It is not possible at a moment for the reason, sketched above, that a phase of experience – primal impression, retention, protention – is only conceivable as an element of a stretch of experience. A succession of momentary experiences is thus needed. On the other hand, it requires the complex protention-primal impression-retention structure for the reason that a mere succession of experiences does not, as both James and Husserl insist, constitute an experience of succession.

On this way of interpreting the Husserlian view, the experience of an event is something that itself must possess duration. As we have seen, Husserl himself insists on this and that the temporal extension of the experience is concurrent with that of the event experienced. Understood in this way, as requiring the temporal extension of conscious experience itself, Husserl seems to have materials with which to reply to the concurrence objection.

In fact, on this interpretation, the Husserlian model is closer to that of some of his critics than is often supposed. For example, Dainton claims that the recognition of concurrence, 'the view that acts and awareness and their contents exactly coincide in time,' forces us to accept that 'awareness is *not* packaged into momentary acts' (2000, p. 166). This amounts, in his view, to a rejection of Husserl's account. Similarly, Phillips claims that 'there are good reasons to reject any theory which attempts to account for temporal experience in terms of our experiencing durations at an instant' (2010, p. 197). Both Dainton and Phillips defend views according to which 'in order to understand the nature of present perceptual experience, one must look beyond the instant' (Phillips, 2010, p. 197). I have been suggesting that, contrary to the way his view is sometimes and perhaps understandably understood, Husserl does just that.

3.5 Immediacy

The above interpretation of Husserl also provides an answer to an objection that Dainton has presented, to the effect that Husserl's model makes the experience of succession less immediate than our experience of simultaneity, something contradicted by the observations, with which we began, of experiencing the movement of the second hand. Dainton claims that, although primal impressions are 'introduced in response to the obvious difference between immediate and represented experience [...] whatever direct awareness we have of phenomenal duration and continuity is located in the retentional matrix, rather than at the level of primal impression' (2000, p. 155), concluding that we are less than immediately aware of change.

In response to this the Husserlian might press on the notion of immediacy, as it is clearly doing much work here and it is less than obvious how it should be interpreted. Whatever immediacy amounts to, however, the charge can be avoided if we interpret Husserl, as I suggested in the previous section we should, as maintaining that the experience of an event (of succession, of change) requires a continuous stream of experience, each moment of which contains protentions, primal impressions and retentions. For, on that picture, it is false to say that awareness of duration is 'located in the retentional matrix'. Rather, the awareness of duration is located in the constant 'movement' of new primal impressions into retentional awareness, 'sinking' ever further into the past. There is nothing less than immediate here and it seems that we have not here met a reason to dismiss the Husserlian picture of our awareness of the passage of time.

4 Conclusion

According to Husserl – and, for somewhat different reasons, Heidegger – the experience of time is the most fundamental of phenomenological issues. It is, arguably, more basic than the experience of things and their properties, investigated in the two previous chapters. But it raises its own puzzles. Specious present views, of the sort offered by James, are problematic in a number of ways. Husserl presents one of the most carefully worked out accounts there is of our awareness of events as spread out in time in an attempt to solve these puzzles. Much work is currently being done on the phenomenology of time, and the question of whether a broadly Husserlian account is acceptable is very much open.

6 Experiencing possibilities

in the weave of the synthetic acts of consciousness there appear at times certain structures that we call imaging consciousnesses. They are born, develop, and disappear according to laws specific to them

Sartre, *The Imaginary*

As I look through my window I see a tree with its characteristic shape and colour, its branches swaying in the breeze. As I close my eyes, I visualise the tree. I picture its branches, still swaying. All of a sudden I imagine it being eaten by an enormous T-Rex. In imagining these things I seem to be aware of possibilities – of possible ways that the tree is and even of a merely possible thing. But what am I doing in imagining these things? More specifically, what am I aware of when I imagine a tree and how should we characterise the form of awareness involved? These questions concerning the phenomenology of imagination open up an aspect of our experiential lives that is at once rich, familiar, and difficult to characterise.

1 Sensory imagination

1.1 'Mental images'

I can, if I like, imagine a tree. Put in another, less philosophically innocent way, I can form a mental image of a tree. It should be pointed out from the very beginning that the terminology of 'mental images' is problematic, and for at least two reasons. First, it is highly suggestive of a particular view of what it is that I do when I imagine something – I 'see' a picture of it in 'my mind's eye'. Such a view of the imagination is, however, open to question and, as we shall see, is rejected by many. Second, it applies only awkwardly to cases of non-visual imagination. For, just as I may imagine a tree as it looks, I may imagine it as it sounds – the rustling of its leaves. We are all familiar with hearing a tune playing 'in one's head', a case of purely auditory imagination. Similarly, I may imagine how the tree feels, smells, or tastes. In thinking about what the objects of imagination are, we must be doubly careful with the term 'mental image'.

1.2 Sensory and suppositional imagination

In the above, the sort of imagination involved can be characterised as 'sensory'. The ways in which I can imagine a tree in some sense mirror the ways in which I may perceive a tree. This is reflected in the fact that we use the term 'visualise' to describe the activity of imagining something as it looks (although colloquial English lacks analogous terms for the other sensory modalities). There is, it seems, an affinity between imagining and perceiving. Understanding exactly how to describe the relation between the two is one of the central tasks of this chapter.

There is another use of the term 'imagine' that lacks this affinity with perception. Thus, I can imagine that the moon is made of cheese, that chess has been solved, or that I am a brain in a vat. None of these, it seems, are particularly closely connected to perceptual experience. This variety of imagining is more akin to supposing something to be the case (e.g. for the sake of argument). Such suppositional imagination is closer to judgement than to perception. Sensory and suppositional imagination differ with respect to how their objects are given. For, so many have found it plausible to say, when we sensorily imagine a tree before us we are, in some sense of the word, *presented* with a tree. When, on the other hand, we suppositionally imagine that there is a tree before us, we are not. This mirrors, at least to a certain extent, the difference between perception and judgement, only the former of which *presents* its object. Of course, in any actual case of imagining, the sensory and the suppositional forms may intertwine. I may imagine, for example, that my visualised tree has roots that extend to the centre of the earth. However, as indicated above, it is the sensory imagination, the form of imagination that in some sense presents its object, that will be our focus.

1.3 Sensory imagination and possibility

The title of this chapter deserves some explanation. What is the connection between the imagination and experiencing possibilities? We have already encountered the imagination in Chapter 1. In his account of the phenomenological method, Husserl describes the 'free variation in imagination (phantasy)' which, on his view, enables us to come to knowledge of the essence of phenomena. For example, when I see a tree it is given to me as a certain shape, size, colour, etc. Each of these I can vary in imagination. Thus, I can imagine a tiny but extraordinarily wide, purple tree. I cannot, however, imagine a tree entirely lacking in shape, size, or colour. The same seems to hold for any experience of something *as a thing*. According to the Husserlian position, we may conclude from this that the visual experience of a thing *as a thing* is necessarily the experience of it as shaped, sized, and coloured.

What would justify this move from the incapacity to imagine a shapeless thing to the claim that an experience of something as a thing must be an experience of it as a shaped thing? On the face of it, the move rests on a connection between imaginability and possibility. More specifically, it seems to rest on something like the assumption that every possible object of perceptual experience is also a possible object of the sensory imagination. One way, then, to describe the sensory imagination is as our way of experiencing possibilities of perceptual experience. This, once more, brings the above-mentioned affinity between sensory imagination and perception to the fore. Of course, sometimes I may imagine things that are not only possible but actual. I imagine the particular tree in my garden, for example. I may even imagine it as it in fact is. But since, of course, anything actual is possible, this is no reason to reject the thought that the imagination is our way of experiencing possibility.

In Chapter 1 I sketched some worries that one might have about this connection between possibility and imagination and I will not rehearse them here. For present purposes, the important points are, first, that the imagination is of central importance to an orthodox conception of Husserlian phenomenological method and, second, that the connection between sensory imagination and perception is of particular significance (for an extended discussion, see Casey, 2000).

2 Sensory imagination as reproduced perception

Perhaps the simplest way to think about the relation between perceptual experience and the sensory imagination is to suppose that imaginings are reproduced perceptions. As Husserl pointed out in an early lecture course on the imagination – which, although not the central focus of this chapter, we shall occasionally have reason to consider – this natural thought is suggested by the phenomena themselves, for '[p]erceptual appearance and phantasy appearance are so closely related to one another, so similar, that they immediately suggest ideas about the relationship of original and image' (Husserl, 1904–5, p. 10; Sartre discusses aspects of Husserl's account of imagination in 1936, Part 4). As we shall see, however, despite its immediacy and historical pedigree this model is problematic.

2.1 Humean impressions and ideas

Hume, in the *Treatise*, introduced the topic of imagination by comparing it with memory and relating both to what he calls 'impressions':

> We find by experience, that when any impression has been present with the mind, it again makes its appearance there as an idea; and this it may do after two different ways: either when in its new appearance it retains a considerable degree of its first vivacity, and is somewhat intermediate betwixt an impression and idea; or when it intirely loses that vivacity, and is a perfect idea. The faculty, by which we repeat our impressions in the first manner, is called the

MEMORY, and the other the IMAGINATION [...] in the imagination the perception is faint and languid, and cannot without difficulty be preserv'd steddy and uniform for any considerable time.

(1739–40, pp. 8–9)

On this picture, when I visualise a tree, I reproduce a, now faint and languid, impression of a tree, the impression in question presumably being a visual experience. As we saw in Chapter 3, on the Humean view what I am aware of in perceptual experience are images, mental entities distinct from worldly Objects. Thus, in the ordinary visual perception of a tree I am, in fact, only indirectly related to a tree, the immediate object of my experience being an image of a tree. With this in mind, and conceiving of the sensory imagination as a reproduced, albeit faint, perception, we can see that the Humean view treats the sensory imagination as the awareness of a mental image.

With respect to visual experience itself, the Humean view strikes many as counter-intuitive. Perception, naïvely understood, puts us in immediate experiential contact with Objects themselves, not images of them. As we saw in Chapter 3, an alternative picture sees perception as an intentional direct-edness to Objects. With respect to the imagination, however, a broadly Humean view may seem to have a great deal to recommend it. For, as we have seen, there is evidently an affinity of some sort between visualisation and visual experience and, more generally, between the sensory imagination and perception – sensory imaginings seem akin to possible perceptions. Both, in some sense, seem to *present* us with objects and are, in this respect, unlike judgement. But, unlike the visual experience of a tree, it is entirely ordinary to enjoy the imaginative experience of a tree with one's eyes closed, or when there are no trees in one's vicinity. What better way to account for these two features of the sensory imagination than to suppose that it involves a perceptual awareness not of a tree but of a mental image of a tree? That is, one may find it plausible that the sensory imagination involves a reproduced perceptual consciousness but with a different, imagistic, object. As plausible as it may at first glance seem, however, and whether we think of perception as a relation to images or as an intentional directedness to Objects, the view of the sensory imagination as a reproduced perceptual experience is subject to a number of objections.

2.2 Imaginary objects: 'the illusion of immanence'

Sartre (1940) refers to the view of imagination as involving the perception of a mental image as 'the illusion of immanence'. It is an illusion, according to Sartre, to suppose that the object of sensory imagination is an immanent entity, existing within the mind, rather than an external, transcendent object. On what grounds does he suppose it to be an illusion? By Sartre's lights, the Humean view is certainly objectionable insofar as it incorporates an implausible imagistic (sense-data) account of perceptual experience. But, as suggested above, one might suppose that this element of the view can be disposed of, retaining only the claim about the object of imagination. What reason does Sartre provide to think that we should reject the account of sensory imagination as the perception of a mental image? Discussing such a view, Sartre writes that it is

> impossible to slip these material portraits into a conscious synthetic structure without destroying the structure, cutting the contacts, stopping the current, breaking the continuity. Consciousness would cease to be transparent to itself; everywhere its unity would be broken by the inadmissible, opaque screens [...] if we accept the illusion of immanence, we are necessarily led to constitute the world of the mind from objects very similar to those of the external world and which, simply, obey different laws.
>
> (1940, p. 6)

There is a great deal going on in this passage and Sartre provides little by way of explanatory discussion. There is a concern here with the unity of consciousness, Sartre's views on which we shall look at in Chapter 7. There is also, however, mention of the transparency, or 'emptiness', of consciousness, something that we have already encountered in Chapter 3. Consciousness, Sartre (1940, p. 4) claims, is empty, nothing but a directedness to its objects. When one reflects upon one's own perceptual experience with the goal of offering a phenomenological description, all that one is presented with are the objects, properties, events that one is perceptually aware of. No further features of perception are given to reflection. In Chapter 3 we saw Sartre wield transparency against the Husserlian notion of *hyle*. In the present context Sartre suggests something similar with respect to mental images. Consciousness is transparent and the supposition that consciousness contains mental images – entities modelled on pictures – is inconsistent with this fact. Further, recalling a concern about Husserlian *hyle*, suppose that I imagine a large green tree: is my mental image large and green? Presumably not. Rather, it is large-ish and green-ish. But without further explanation we are at a loss as to what these terms mean. As Sartre here puts it, the view leads us 'to constitute the world of the mind from objects very similar to those of the external world and which, simply, obey different laws'. Reflection on the experience of imagining reveals no mental images which are, in any case, dubiously coherent. Rather, Sartre concludes:

> whether I perceive or imagine this straw-bottomed chair on which I sit, it always remains outside of consciousness. In both cases it is there, *in* space, in that room, in front of the desk. Now – this is, above all, what reflection teaches us – whether I perceive or imagine that chair, the object of my perception and that of my image are identical: it is that straw-bottomed chair on

which I sit. It is simply that consciousness is *related* to this same chair in two different ways.

(1940, p. 7)

Sartre, then, denies that when I visualise a tree I perceive a mental image of a tree. Rather, I am aware of a transcendent tree. Of course, in our example, the transcendent tree is an actually existing tree (as is Sartre's straw-bottomed chair). Many, if not most, instances of sensory imagination are not like this, however. In imagining the tree being eaten by an enormous T-Rex I am imagining a merely possible object. This does not mean, though, that the object is not transcendent. For Sartre's view is, of course, that the sensory imagination is an intentional awareness of transcendent Objects and, as we saw in Chapters 2 and 3, this view is taken by its proponents to allow for cases in which the intentional objects of intentional states do not actually exist. It does not matter, for the intentionalist, whether or not the object exists, whether or not we imagine some particular existent tree or just a tree in general. The imagination gives us possibilities of perception, but they need not be actualities.

Given Sartre's rejection of mental images, it is perhaps unfortunate that he continues to use the term 'image' in his own discussion to refer to 'a certain way in which the object appears to consciousness' (1940, p. 7), his slogan being that 'the image is a consciousness'. In his view, mental images are not the *objects* that we are aware of but our *awareness* of them. This use of the term 'image' is, however, apt to mislead and it seems to me that the simplest way to describe Sartre's critique of the illusion of immanence is to say that he denies that the imagination involves mental images at all (for a recent defence of mental images, see Kind, 2001).

As his talk of 'two different ways' of being aware of objects suggests, the issue of transparency is not Sartre's only complaint with the account of sensory imagination as the perception of a mental image. In fact, he provides a number of reasons to think that the sort of awareness that we have of the objects of imagination is not to be straightforwardly modelled on perception, as he seems to assume that it would have to be if one were to accept the 'illusion of immanence'. That is, not only is the sensory imagination not a perception of a *mental image*, it is not a *perception* at all.

2.3 Imaginary experience: quasi-observation, positing, and spontaneity

Perceptual experience, according to Sartre, is an observational form of awareness that posits its object as existent and is, in a certain sense, passive. Sensory imagination, on the other hand, is a quasi-observational form of awareness that does not posit its object as existent and which is, in the relevant sense, spontaneous. Sartre's defence of these differences between perceiving and imagining plays two roles in his overall dialectic. First, it further supports his rejection of the view that sensory imagining is reproduced perception, for it is quite unlike perceptual awareness in its structural features. Second, given Sartre's insistence that the objects of sensory imagination are identical to the objects of perceptual awareness, i.e. are transcendent, an account is needed of how imagination and perception differ.

Quasi-observation

The imagination is like perception in some respects and like thinking in others. Like perception, when I imagine a tree it is presented to me from a particular perspective – I am aware only of its facing side. This feature of perceptual experience – the fact that 'the object of perception constantly overflows its object' (Sartre, 1940, p. 10) – has been a key theme in the preceding chapters. Perception, as Sartre puts this point, is observational.

Thinking of an object does not share this feature. In thinking of a cube, 'I think of its six sides and its eight angles at the same time. I am at the centre of my idea, I apprehend its entirety in one glance' (Sartre, 1940, p. 8). As with an object thought about, an imagined object is presented to me in such a way that there is not 'more to learn'. In this way the sensory imagination differs from perception. Despite the fact that they are given perspectivally, there is never any more to my mental images than I put into them. As Sartre puts it, '[w]e are, indeed, placed in the attitude of observation, but it is an observation that does not teach anything' (1940, p. 10). The sensory imagination is, in Sartre's terminology, 'quasi-observational', the imagined object having an 'essential poverty' (1940, p. 9).

Given that, in perception, I am aware of the object as 'having more to show', if imagining were a perceptual awareness of an inner mental image, the perceived image would exhibit this characteristic. Yet, according to Sartre, it does not. The phenomenon of quasi-observation, then, gives us reason to doubt the thought that imaginary experiences are reproduced perceptual experiences, since they lack a feature common to all perceptual experiences, that of being observational in Sartre's sense.

This may, however, be challenged on the grounds that it is, in fact, possible to learn something *via* the imagination. One might, for example, visualise carrying a sofa into a room and come to learn that it will not fit through the door. Examples such as this, one might suppose, provide reason to think that one does, contrary to Sartre's claim, inspect one's mental images. Since such examples seem entirely commonplace, isn't Sartre wrong to claim that the imagination 'does not teach anything'?

Whilst examples of this sort are easy to come by, it is not clear that they really challenge Sartre's account of quasi-observation. What the sofa example teaches pertains to the relation between two already known features – the size of the sofa and the width of the door-frame – and this is not what the phenomenon of observation really concerns. Rather, perception is observational in the sense that its objects 'overflow' one's current awareness of them. When I see the sofa, it looks such that were I to get a closer look, or turn it around, I would have a better grip on some of its features. Sartre's claim is that this is not true of the imagination. To illustrate, as I look at the tree beyond my window, I can count its major branches (there are six). However, if I imagine a tree, I cannot do the same. If the imagined tree has six branches, this is because I made it have six branches: 'one can never learn from an image what one does not know already' (1940, p. 10).

Positing

Not only is perceptual experience perspectival, it is also a form of awareness in which an object – be it thing, property, or event – is given to me as existing here and now, as *present* in both its spatial and temporal senses. Thus, when I see the tree beyond my window it seems to exist here and now. The same can be said of its shape and of its gentle swaying. Each are present, and this is a familiar fact from our discussion of the natural attitude that, according to Husserl, must be bracketed for the purpose of phenomenological description. The objects of the sensory imagination, however, lack this feature. In imagining a tree, whilst there is a sense in which I seem to be presented with a tree, the imagined tree is not given to me as existing here and now. Indeed, in many cases it is not given to me as existing at all, as when I imagine the tree being eaten by a T-Rex.

Sartre speaks of the way that in perception an object is given as existing as the way in which it 'posits' its object, a notion that he takes over from Husserl (1913, §117), and which he sometimes suggests involves belief (1940, pp. 12 and 14). That is, when I perceive a tree, I believe that it exists here and now. This, however, cannot be right, since the phenomenon seems to persist in the absence of belief. If I were told by a normally reliable source that my current visual experience is hallucinatory, I would cease to believe that the tree I seem to see exists (here and now). But surely it would still *seem* to do so. If I were to take my experience at face value, I would believe it to exist. Thus, there is reason to think that the fact that perception 'posits its object as existing' is not a matter of its including or being invariably accompanied by a relevant belief. More plausibly, positing is a feature of perceptual experience itself, according to which it 'asserts' the actuality of the object (see Carman's account of positing as analogous to 'illocutionary force' in 2003, pp. 67–71). Thus, just as the tree is given as possessing a particular shape and size, so it is given, perceptually, as existing here and now. Indeed, in discussion of the way in which the imagination posits its object, Sartre suggests something similar, claiming that 'the positional act is constitutive of the image consciousness' (1940, p. 13).

How, then, does the sensory imagination posit its object? Not, Sartre tells us, as perception does, as existing here and now. Rather, the act can take four and only four forms: it can posit the object as non-existent, or as absent, or as existing elsewhere; it can also 'neutralize' itself, which is to say not posit is object as existent.

(1940, p. 12)

These ways in which the imagination posits is object share a negative character – in various ways the object of imagination is given but *not* as actually existing here and now. This variety is captured in Sartre's claim that the imagination 'gives its object as a nothingness of being' (1940, p. 13). In imagining the tree, I posit it as existing elsewhere, in imagining the T-Rex, I posit it as non-actual (but possible), and so on. This account echoes that of Husserl, who writes that 'the *phantasm*, the sensuous content of phantasy, gives itself as not present. It defends itself against the demand that it be taken as present; from the beginning it carries with it the characteristic of irreality' (Husserl, 1904–5, p. 87).

As with the case of quasi-observation, this difference in the way that perception and imagination posit their objects provides philosophical mileage. Given that, in perception, I posit the object as actually existing, if imagining were a perceptual awareness of an inner, mental image, the perceived image would exhibit this characteristic. Yet it does not. So the phenomenon of positing as a nothingness of being gives us reason to doubt the thought that imaginary experiences are reproduced perceptual experiences since, once more, they lack a feature common to all perceptual experiences.

Spontaneity

Perceptual experience seems to the perceiver to be an awareness of a world that is there anyway. In opening my eyes, I seem not to create objects but to become open to a world that exists independently of my view of it. Sartre describes this as the 'passivity' of perceptual experience. Of course, perception is active in some sense – I control my perceptual experiences by controlling the direction in which my body, head, and eyes are pointing. However, it is passive in a more significant sense – I cannot, at will, perceive just any object I please, nor are the objects I do perceive given to me as the creations of that perceiving. In this respect, the sensory imagination is markedly different. Not only can I imagine any object I wish, those objects are presented to me as created by that very act. Employing the notion of pre-reflective self-awareness that we will consider in detail in Chapter 7, Sartre marks this distinction in the following way:

A perceptual consciousness appears to itself as passive. On the other hand, an imaging consciousness gives itself to itself as an imaging consciousness, which is to say as a spontaneity that produces and conserves the object as imagined. It is a kind of indefinable counterpart to the fact that the object gives itself as a nothingness.

(1940, p. 15)

The object that I sensorily imagine is given as 'produced and conserved' by an act of imagination that is itself given as productive, as spontaneous. As Sartre suggests, this feature of the imagination is connected with that of positing. In being given, in this way, as creating and preserving, the object is thereby given as created and conserved. Thus it is given as a nothingness of being, a possibility, rather than as actually existing here and now. But it also connects with quasi-observation, since it is the spontaneous character of the imagination that accounts for the fact that there is nothing in the object as imagined beyond what the imaginer puts there. Its 'essential poverty' is explained by its being presented as the mere shadow of a spontaneous act on my part.

The spontaneity of the sensory imagination can also be brought out by way of its connection to suppositional imagining. Earlier, I suggested that an intentionalist such as Sartre need not be overly concerned with the fact that some imagined objects do not exist. Sometimes we imagine an actual particular tree, sometimes we might imagine a merely possible tree in general. But, considering the former sorts of case, what makes it such that my imagining is an imagining of a particular tree? In the case of an analogous question asked of perceptual experience one may be tempted to offer a causal answer: my current perceptual experience is an experience of *that* particular tree because it has been caused by that particular tree. Whether or not that is right, in the case of the imagination the answer is surely that I *stipulate* which tree it is that I am imagining. That is, I sensorily imagine a tree and simultaneously suppositionally imagine it to be the tree beyond my window, or Arbol del Tule, or whatever. Thus, the identity of the imagined object is determined spontaneously.

It will be objected that not all cases of imagination are spontaneous. For example, in daydreams my imagination may run wild in such a way that it seems inappropriate to describe the things, properties, and events imagined as 'put there' by me. Similarly, it is obvious enough that the imagination is not subject to the will in a straightforward way. However hard I try, and no matter how many times I have seen it, I may be unable to visualise the house in which I was raised. Conversely, I may be unable to prevent myself from compulsively imagining some gory horror movie scene, no matter how unpleasant. In response, Sartre can point out that spontaneity, as he intends it, is not at all the same as being subject to the will, although the two bear a close relation. Thus, something can be spontaneous without necessarily being subject to the will. Even when compulsively imagining, the gory scene is given as a non-actual object that is created and sustained by my own irresistible act of imagining. This is the sense in which the imagination is spontaneous and the sense that matters for Sartre's purposes.

A perhaps more pressing worry concerns the consistency of spontaneity, so characterised, with Sartre's overarching claim that the objects of sensory imagination are transcendent; that the tree I imagine is identical to the tree that I see (Hannay, 1971, Ch. 4). Perceptual experience is an openness onto a world that exists anyway. But if the sensory imagination is a distinct way of being conscious of that very same world, Sartre would surely contradict himself if he were to claim that the object of the sensory imagination is created by one's act of imagining it. In response to this subtle objection, Sartre can point out that he does not claim that the object of sensory imagination is created by the act of imagining it, but that it is *given as* so created and conserved. Imaginative consciousness 'gives itself to itself as an imaging consciousness' (1940, p. 14).

But this response merely leads to another objection, which is that the object of my imaginative experience when I imagine the particular tree beyond my window is precisely *not* given as created by my act of imagining it. Rather, just as with perceptual experience, it is given as there anyway. The response to this second worry is to distinguish, as Sartre rather carefully does, between the thing itself (the Object) and the object-as-imagined. The Object imagined is identical to the Object seen. But the object-as-imagined is the object *as presented in imaginary experience* or, alternatively, *the way in which the object is imagined*. The latter is what Sartre claims is given as created by the act of imagination, the former is not. On this interpretation, then, that which is given as created is not the Object, but rather the way that the Object is given. Sartre's claim is that the tree is given in a certain way and its being given in just this way is, and is given as, a shadow of my imagining it. This, so he claims, is not the case with perception. In that case, the tree's being given in a certain way is a consequence of the way the tree is, and the relation in which I stand to it.

As with quasi-observation, this claimed difference between perception and imagination has significance for Sartre. Given the passivity of perception, if imagining were a reproduced perceptual awareness of either an inner or outer object, the imagination would itself be given as passive. The fact that it is not is further reason to reject the picture of imagination as the perception of a mental image.

3 Sensory imagination as seeing-in

Sartre has thus far presented a view according to which the object of my visualising is, just as with vision, the tree itself. Visualising is distinguished from perception, however, in being spontaneous, in positing its object as a nothingness, and in being only quasi-observational. As Sartre recognises, though, this is inadequate as a complete account of the sensory imagination. For, as we noted earlier, when I imagine a tree there is a sense in which I am presented with a tree. This, at least in part, is what motivated the suggestion that we understand the sensory imagination as the perception of inner, mental images. The current account does not seem to capture the sense in which visualising is presentational. Of course, the characterisation of quasi-observation as including the claim that a visualised tree is presented perspectivally *presupposes* the presentational character of visualising, but it does not tell us in what such presentational character consists. So Sartre owes an account of this. But if, due to such structural differences between imagining and perceiving, visualising a tree cannot be modelled on seeing a tree, what would be a good way to understand it? Sartre's answer is that we should understand visualising a tree not on the model of experiencing a tree, but on the model of experiencing a tree *in a picture of a tree*.

3.1 Experiencing pictures

If I see a good picture of my tree, whether it is a drawing, painting, or photograph, I am aware of more than one thing. Most obviously, I am aware of the picture-object itself: the canvas and paint, the card and ink, or whatever. But, equally, I also see a tree in the picture. By way of my looking at the picture, a tree becomes present to me in a particular way. More than this, however, through the picture I see a particular tree, my tree. This notion of seeing-in, of seeing an object in another, provides Sartre with a model for the sensory imagination:

> Mental images, caricatures, photos are so many species of the same genus [...]. These various cases all act to 'make present' an object. This object is not there, and we know that it is not there. We therefore find, in the first place, an intention directed at an absent object. But this intention is not empty: it directs itself through a content, which is not just any content, but which, in itself, must present some analogy with the object in question.
>
> (1940, p. 19)

To appreciate what is going on in this passage, we must briefly return to Husserl's account of perceptual experience. Recall from Chapter 3 that Husserl, and following him Heidegger, characterise a perceptual experience as one in which an intended object is 'bodily present'. On Husserl's account, being bodily present is a matter of being presented in an experience that is intuitively fulfilled by way of non-intentional, sensory *hyle* or 'content'. Recall, this use of 'content' is not to be confused with the contemporary notion of representational content but is in fact contrasted with it. It is the sensational 'content' to the intentional 'form'. These sensory features of experience are not themselves the objects of perceptual experience but are 'animated' by intentions, thereby coming to serve as representations of the object perceived. In this way perceptual experience is distinguished from the empty intending of judgement.

In the above we see Sartre claim that when I see a picture of my tree, that experience is not empty but 'directs itself through a content'. Another way of putting this would be to say that my experience does not have the tree as its intentional object in the same way as does a judgement about the tree. I do not see a picture-object and merely judge it to be a picture of the tree in question. Rather, I am presented with the tree. In the experience of a depicted tree, however, the tree does not become bodily present as though it were genuinely perceived. Rather, it is experienced as absent, as possible, as a nothingness. Looking at a picture of a tree is not just like looking at a tree. I do not *see* (or otherwise perceive) the depicted tree but it is nevertheless presented to me. It is, as it can be rather paradoxically put, presented as absent.

Sartre's use of the Husserlian notion of 'content' may seem surprising given that, as discussed in Chapter 3, he rejects Husserlian *hyle* as incoherent, claiming instead that conscious experience is empty. If, as Sartre claims, we can and should dispense with *hyle* – non-representational sensational features of experience – how can we understand his above description of the experience of a picture? The answer to this question is that Sartre sees the picture-object itself – a non-psychological, transcendent entity – as playing the role of 'content'. Thus, in experiencing a picture of a tree, we do not animate, or construe, perceptual sensations as appearances of a tree, rather we animate the paint on the canvas as one. The picture-object is that by means of which the tree is presented as absent (Sartre rather confusingly uses 'matter' where 'content' would be more appropriate): 'The photo is no longer a concrete object that provides me with perception: it serves as matter [content] for the image' (1940, p. 21).

> Pierre, on the one hand, can be far from his portrait […]; but it is exactly this 'object far from us' that we aim at. But, on the other hand, all the physical qualities are there, before us. The object is posited as absent, but the impression is present.
>
> (1940, p. 23)

On this view, the features of the picture that serve as the 'content' of my experience of the tree, or of Pierre, are no longer given as themselves, 'as a spot of colour on a canvas', but rather 'the coloured spots on the picture give themselves to the eyes as a forehead, as lips' (1940, p. 23). Although I see the spots, I see them not as spots but as parts of that which is depicted.

There is a great deal more that can be said about the notion of seeing-in (cf. Hopkins, 1998). Already, however, we have most of the material with which to appreciate Sartre's account of sensory imagination. For just as, in experiencing a picture of a tree, I experience an absent tree *via* a present picture-object, so too when imagining a tree I experience an absent tree *via* a presence.

3.2 Experiencing the imaginary

There is a familiar form of sensory imagination, imagining-as, that we have not yet mentioned but which is helpful to consider in making the transition from the experience of pictures back to the sort of visualisation with which we began. If I look at the bare branches of the tree before me, I may imagine them to be blood vessels within an invisible brain. This is not simply a case of suppositional imagining, nor do I pretend that the branches are blood vessels. Rather, my imaginative act has a visual character. The mass of branches in some sense takes on the look of a system of blood vessels. Sartre's account of experiencing pictures can help us to understand this. In imagining the branches to be blood vessels I no longer see them as branches but as blood vessels. In adopting such an imaginative attitude, my experience is intentionally directed towards an absent and merely possible brain by way of a present perceptual awareness of an object that is analogous to it. I experience a brain in the tree.

Sartre's model works well for such cases for the reason that they manifestly involve some transcendent object of which we are aware and that can function as the 'content' by which we imagine something. But the case with which we began, in which I close my eyes and imagine the tree being swallowed by a T-Rex, involves no such object. Although there is nothing that I can see, I am nevertheless presented with a tree. At first sight it seems difficult to see how Sartre could account for such paradigmatic cases of 'pure' sensory imagining.

A first hint is given when we notice that in the following general statement of his view, he allows that the 'content' for an imaginative awareness can be either physical or mental ('psychic'),

> the image is an act that aims in its corporeality at an absent or nonexistent object, through a physical or psychic content that is given not as itself but in the capacity of 'analogical *representative*' of the object aimed at.
>
> (1940, p. 20)

In cases of pure sensory imagining, Sartre claims, the 'content' (or 'analogon') is a mental entity. But this should not be seen as a return to the mental image view (cf. McCulloch, 1994, Ch. 5), since 'one does not see a mental image' (Sartre, 1940, p. 52). Nor is it a return to the Husserlian *hyle*, since whilst the analogon is mental it is nevertheless transcendent to the imagining consciousness. As Sartre puts it:

> This necessity for the matter [content] of the mental image to be already constituted as an object for consciousness, I call the *transcendence* of the representative. But transcendence does not mean externality: it is the represented thing that is external, not its mental 'analogon'.
>
> (1940, p. 53)

We are looking, then, for some mental entity that may serve as the analogon for my imagining the tree. It is at this point, however, that Sartre hits a dead end, claiming that 'reflective description does not directly tell us anything concerning the representative matter [content] of the mental image,' and that 'we are reduced to conjectures' (1940, p. 53). In Sartre's opinion, then, this signals the limits of phenomenological description. A purely phenomenological account of the sensory imagination can tell us that it has the structure of seeing-in and that, therefore, there must be some representative 'content'. What it cannot do is tell us what that content is.

Sartre does, however, offer a lengthy discussion of what such 'contents' may be. For present purposes, we may sidestep the detail, simply reporting that Sartre offers two options. First, contents may be constituted by affective experiences (emotional feelings). In thinking of my childhood home, I may feel a wave of nostalgia. This feeling, itself a mental entity, may serve as the analogon for an act of imagining that place, since 'in relation to this affective object, I find myself in the attitude of quasi-observation' (1940, p. 72). Second, and perhaps more commonly, the role of the analogon may be played by bodily sensations experienced as I move (kinaesthetic sensations) and, in particular, as I let my eyes trace the outline of the imagined thing. Sartre expends a great deal of effort in attempting to explain how a series of *non-visual* bodily sensations may serve as an analogon for an instance of *visual* imagining, concluding that

> if we form an image of an object, the kinaesthetic impressions that accompany certain contractions, certain voluntary movements of the organs, can always serve as substitutes of a visual form. But this visual form will now have a wider meaning: it could be the form of my fist, of an inkpot, of a letter of the alphabet; in brief, the form of an object.
>
> (1940, p. 81)

This, then, is Sartre's account of pure sensory imagination. It is the spontaneous and quasi-observational awareness of a transcendent object which is presented *via* present affective or kinaesthetic sensations, but itself posited as a nothingness.

3.3 Is there a mental analogon?

The most obvious way in which to reject Sartre's view is simply to deny that, in cases of pure imagining, we are aware of any mental analogon. This objection is put forcefully by Hopkins, who writes that

> Visualizing does not have the structure of seeing-in because, from the point of view of the subject at least, visualizing does not have any structure at all. All the visualizer, *qua* engaged in visualizing, is aware of is the visualized object and the properties it is visualized as having.
>
> (1998, p. 166)

In paradigmatic cases of seeing-in, we are aware of more than one thing – the canvas, an image of a tree, and that particular tree – however, in cases of pure visualising we are aware, so the objection runs, only of the last of these. But this is inconsistent with Sartre's account, which requires us to be aware of the analogon not as itself but as a representative of that which is imagined. If we are not aware of it at all, then we are surely not aware of it *as* anything. Thus, insofar as the phenomenology is concerned, visualising cannot be modelled on seeing-in. Indeed, this objection might be supported by Sartre's own acceptance that the nature of the analogon cannot be determined by reflective description but must be conjectured. Husserl took such a view to be phenomenologically supported, writing that

> In phantasy we do not have anything 'present' and in this sense we do not have an image object. In *clear phantasy*, we experience phantasms and objectifying apprehensions, which do not constitute something standing before us as present that would have to function first of all as the bearer of an image consciousness.
>
> (1904–5, p. 86)

Sartre may respond to this objection by pointing out that those mental entities that he considers to be potential 'contents' for imaginary experience are not unconscious. Affective and kinaesthetic sensations are themselves conscious experiences of which one can be aware. Since, insofar as we are conscious, we are always aware of *something* there will always be some experience or other that may be taken as an analogon. But this is not yet an adequate response since, as Sartre himself recognises, not just anything may serve as an analogon for anything else. It must be *analogous* to it, and that means that the two must bear some resemblance. Simply being told, without further explanation, that kinaesthetic sensations are the analogon for imagining a tree 'is a little like being told that goats function as analogical substitutes for seaplanes' (Sartre, 1940, p. 76).

At this point one might insist on the importance of eye movements. For, in imagining a tree, it is difficult not to let one's eyes trace its outline. What they trace, a form, is precisely the sort of thing that is available not only to tactile and kinaesthetic awareness but also to vision – one can both feel and see shapes. Thus, there is perhaps some sense in which one's kinaesthetic sensations resemble the object imagined. Such a response, however, has significant limitations. We can easily imagine a green tree and it is far from clear what eye, or other bodily, movements could resemble the colour of an object (although see Noë, 2004, Ch. 4). It seems clear that more needs to be said on this score.

3.4 Is there a need for a mental analogon?

A more radical objection challenges the whole structure of Sartre's account, questioning the necessity of positing an analogon at all. On Sartre's view, the fact that visualising has the structure of seeing-in means that there must be something that the visualised is experienced in. This is the analogon. That visualising has the same structure as seeing-in is motivated, at least in part, by the need to account for its presentational character. But one might deny that the presentational character of visualising is best accounted for in this way. If an alternative to the seeing-in model can be found, we will do away with the need to find an analogon for sensory imagining. In what follows I will briefly sketch one suggestion along such lines.

In his discussion of the object of sensory imaginings, Sartre considers two options: the transcendent object itself and a mental image of it. Thus, in visualising a tree, I am aware either of a tree or a mental image of a tree. But there is another possibility that has been defended in recent literature. Peacocke (1985) defends the view that the objects of sensory imagination are possible experiences. That is, visualising a tree is imagining seeing a tree, auditorily sensorily imagining is imagining hearing, and so on. At first glance one might suppose that this is a return to the illusion of immanence, with visualising construed as the perception of an inner image. But this would be a mistake. The view, rather, is that the *transcendent* object of imagining is an experience. This object is an experience that one is not currently having, it is merely possible (remember that we have been assuming throughout that it is possible to imagine things that do not in fact exist).

This view promises to explain the presentational character of visualising without appealing to seeing-in. For, so the thought goes, visualising a tree inherits the perspectival, presentational character of seeing. Thus, just as with certain views of visual experience itself, one might hope that this picture of sensory imagination exhaustively explains its phenomenology, in particular its presentational aspect, by means of the object towards which one is inten-tionally directed. There is simply no need, on this view, to posit anything with the structure of seeing-in.

Defenders of such a view, however, must walk a delicate line. On the one hand, they must suppose, and ideally explain how it is, that visualisation is a form of experience that inherits its presentational character from that of its objects. Only in this way will it be adequate to the phenomenology. On the other hand, it must remain consistent with the phenomenological differences between seeing and visualising – observation versus quasi-observation, etc. – that we have already considered. It is not obvious from this bare sketch how it could do both. Indeed, Hopkins (1998, p. 192) claims that it fails to explain the relevant inheritance. Whether this is a fair criticism will depend on how the view is fleshed out. It is fair to say, however, that despite its own drawbacks, this is a problem that the Sartrean view does not face. On Sartre's view, the perspectival, presentational character of visualisation is explained by its exemplifying the structure of seeing-in. Once again, more needs to be said if we are to settle on a view of the nature of the sensory imagination.

4 Conclusion

In the sensory imagination we are aware of possibilities. But exactly what the objects of such imaginings are is a thorny question. It is perennially tempting to suppose that in visualising a tree I see an image of a tree 'in my mind's eye'. Yet such a view is, as Sartre points out, open to serious objections. But if, as Sartre has it, the objects of the sensory imagination are ordinary (albeit often merely possible), transcendent Objects, an account is required of how the imagination differs from perceptual experience in the relevant modality. Sartre's account is both ingenious, problematic, and far from the last word.

7 Experiencing oneself

As well as being aware of the surrounding world of things, properties, and events, it would seem that I am also aware of myself and my own experiences of those things. In seeing a tree, I am aware not only of the tree but also of my experience of it and of myself as having that experience. Indeed, on Husserl's view, the role of the phenomenological reduction is precisely to enable accurate phenomenological description by reflectively focusing the attention on *experience* rather than the things experienced. As such, we have been implicitly employing a reflective method throughout the previous investigations. But what exactly is reflection? How are its objects (experiences) given? And is there really a distinctive form of awareness of oneself involved in the reflective awareness of my own experience? If so, what form does it take? These are among the most difficult philosophical questions and there have been a great many attempts to answer them. Beginning with the issue of awareness of oneself, and to get a sense of the philosophical landscape in which the phenomenological claims of Husserl, Sartre, and others are situated, we should begin with Hume's infamous denial of self-awareness.

1 Self-awareness

1.1 *Hume vs. Kant*

For my part, when I enter most intimately into what I call *myself*, I always stumble on some particular perception or other, of heat or cold, light or shade, love or hatred, pain or pleasure. I never can catch *myself* at any time without a perception, and never can observe anything but the perception [...]. If anyone upon serious and unprejudic'd reflexion, thinks he has a different notion of *himself*, I must confess I can no longer reason with him. All I can allow him is, that he may be in the right as well as I, and that we are essentially different in this particular. He may, perhaps, perceive something

simple and continu'd, which he calls *himself*; tho' I am certain there is no
such principle in me.

(Hume, 1739–40, p. 252)

Hume is here claiming that reflection upon one's own experience does not
reveal a continuously existing self. On the contrary, reflection reveals nothing
but an ever-changing stream of mental states or events. In Humean terms,
there is no *impression* of the self and, as a consequence of his empiricist claim
that every *idea* be preceded by an impression, the idea that we have of ourselves
is rendered problematic.

However, as Hume saw things, this left him with a problem to which he
could not see the answer: 'all my hopes vanish when I come to explain the
principles, that unite our successive perceptions in our thought or consciousness'
(1739–40, pp. 635–6). This problem concerns the unity of conscious experience,
an issue which has long been closely connected to that of self-awareness.

When I see the tree beyond my window, I also see the grey sky in the
background. I experience these things *together*. In fact, I appear to be simultaneously aware of a large number of things, and not just visually but through
the various sensory modalities, all from the perspective of a single unified
consciousness. Additionally, as we saw in Chapter 5, such unity is apparent
over time. That is, not only do I experience a number of simultaneous things
in some sense *together*, in experiencing duration I also experience a number of
non-simultaneous things *together*.

In virtue of what, one might ask, is this multiplicity of experiences unified
within a single stream of experience? This is the question of what accounts for
the unity of experience. Furthermore, we can ask, in virtue of what is *my*
unified stream of experience differentiated from that unified multiplicity of
experiences that *you* enjoy? What accounts, that is, for the fact that some
experiences are mine and others are yours? This is the question of what
accounts for the individuality of experience.

It is natural to answer these questions by appeal to a notion of *ownership*.
That is, one might claim that I bear the ownership relation to each of my
experiences, and you bear the relation of ownership to each of yours, and so
on. This, it might be thought, answers the individuality question. Furthermore,
not only do I own my experiences, I am experientially aware of that ownership
relation. That is, I am aware of each of my own experiences *as mine* and, it
might be maintained, I am aware of each pair of my own experiences *as
together mine*. This would be an account of the unity of experience in terms of
self-awareness.

However if, with Hume, we find the idea of the self problematic, we will
surely find the idea of ownership problematic. For the notion of ownership
presupposes that of an owner, and what is an owner of experiences if not a
self? The Humean, then, will be unsatisfied with this account of the unity and
individuality of experience in terms of ownership and self-awareness.

Hume's challenge to locate the self amongst the stream of experience was picked up by Kant, who claimed that 'this identity of the subject, of which I can become conscious in every representation, does not concern the intuition of it, through which it is given as object' (Kant, 1781/1787, B408). One way of reading Kant's claim here is that whilst, just as Hume claimed, reflection does not present the self *as an object*, there is nevertheless some form of awareness of the self *as subject* and as identical across the stream of experience. What this form of awareness might amount to, however, is less than clear.

Kant's view of these matters is further complicated by the fact that he disagrees with Hume concerning the legitimacy of the concept of the self. His solution to the two problems of the unity of experience is, as suggested above, that diverse experiences are unified by the synthesising activity of the self:

> The thought that these representations given in intuition all together belong **to me** means, accordingly, the same as that I unite them in a self-consciousness, or at least can unite them therein [...] for otherwise I would have as multicoloured, diverse a self as I have representations of which I am conscious.
>
> (1781/1787, B143)

Thus, Kant requires that the notion of the self as unifier of experience be legitimate. The reason that Kant can allow this despite the lack of an experience (intuition, Humean impression) of the self as an object is that he does not accept the empiricism that drives Hume's account. On the Kantian view, it is legitimate to appeal to a self that unifies experience since such a thing is a condition of the possibility of experience. Without such a unifying self, experience would not be possible, therefore the concept is legitimate. The self, on this account, is *transcendental*, it is introduced as a condition of the possibility of experience, this move being one of the distinctive features of Kantian transcendental philosophy. Thus, while Hume and Kant agree that reflection reveals no awareness of a self as *an object*, they seemingly differ with respect to both whether there is some form of awareness of the self as an identical subject, and also as to whether we may appeal to the concept of the self and self-awareness in an explanation of the unity and individuality of experience.

1.2 Husserl vs. Husserl

Husserl's views on self-awareness and its relation to the unity and individuality of experience evolved over his philosophical career. In *Logical Investigations* he argued against the Neo-Kantian position of Natorp, according to which the self is the '*subjective centre of relation* for all contents in my consciousness' but which 'cannot itself be a content, and resembles nothing that could be a content of consciousness' (quoted in Husserl, 1900–1, Vol. 2, p. 91).

Husserl echoes Hume in claiming to be 'unable to find this ego' (1900–1, Vol. 2, p. 92). Of course, it may be replied that this is precisely because it 'cannot itself be a content' and, as Natorp goes on to say, 'we have ceased to think of it as an ego, if we think of it as an object' (1900–1, Vol. 2, p. 92). But Husserl is unimpressed by this move since, by his lights, either we are aware of the 'pure ego' or we are not. If we are, then it is an object for consciousness, and if we are not then the Neo-Kantian position is without foundation.

Of course, Husserl does accept that, in reflection, we are aware of our own experiences and this, he thinks, amounts to a way of being aware of ourselves.

> the matter is quite clear: there are acts 'trained upon' the character of acts in which something appears, or there are acts trained upon the empirical ego and its relation to the object. The phenomenological kernel of the empirical ego here consists of acts which bring objects to its notice, acts in which the ego directs itself to the appropriate object.
>
> (1900–1, Vol. 2, p. 92)

Husserl claims that an analogy can be drawn between the way in which we are aware of things one-sidedly and the way in which we are aware of our 'empirical ego', the totality of my bodily and mental features. In reflection, I am aware of an experience 'which stands in the same sort of relation to the mental ego as the side of a perceived external thing open to perception stands to the whole thing' (1900–1, Vol. 2, p. 92). Just as, in visual perception, a tree is given through adumbrations with a combination of intuitively full aspects and empty, co-given aspects, one's 'empirical ego' is given to oneself in just the same way. Those experiences upon which my reflection is currently directed are given with intuitive fullness, others are merely co-given. As such, the 'empirical ego' is bodily present and, as Husserl puts it, '[w]e perceive the ego, just as we perceive an external thing' (1900–1, Vol. 2, p. 92), a claim later repeated by Sartre.

At the time of *Logical Investigations*, then, Husserl adopted a Humean scepticism about the Kantian pure ego, but did not concur with Hume that when reflecting one 'never can observe anything but the perception [experience]'. Rather, reflection reveals the empirical ego, *through* one's experiences, in much the same way that perception reveals the thing, *through* its presented aspects. Further, whilst Husserl, like Hume, denied that reflection presents us with a pure self, unlike Hume, he did not appear to find overly problematic any resulting questions concerning unity or individuality.

Later, however, in *Ideas I* and *II*, Husserl presents a notably different view. In *Ideas I* he claims that whilst it is true that the pure ego is neither 'one mental process among others nor strictly a part of a mental process', it nevertheless 'seems to be there continually, indeed necessarily' (1913, p. 132). In defence of this claim that I am necessarily aware of the pure ego, Husserl presents the familiar Kantian thought that all mental processes [...] as belonging to the *one* stream of mental processes which is mine, *must* admit of becoming converted into actional cogitationes [i.e. in the form of 'I think'][...]. In Kant's words, *'The "I think" must be capable of accompanying all my presentations.'*

(1913, pp. 132–3)

Thus, Husserl seemingly offers an account of the unity of experience that appeals to the self functioning transcendentally, as a condition of the possibility of experience.

But what is the 'pure ego' supposed to be and how does it differ from the 'empirical ego', awareness of which Husserl had endorsed in *Logical Investigations*? In *Ideas I*, Husserl says little, maintaining somewhat paradoxically that it is 'a *transcendency within immanency*' (1913, p. 133). That is, it is not an experience (so it is transcendent) nor merely the object of an intentional state (so it is immanent). In *Ideas II*, however, he is more forthcoming, claiming that the pure ego is

> by no means whatsoever something mysterious and mystical. I take myself as the pure ego insofar as I take myself purely as that which, in perception, is directed to the perceived, in knowing to the known, in phantasizing to the phantasized, in logical thinking to the thought, in valuing to the valued, in willing to the willed. In the accomplishment of each act there lies a ray of directedness I cannot describe otherwise than by saying it takes its point of departure in the 'Ego,' which evidently thereby remains undivided and numerically identical.
>
> (1952, pp. 103–4)

There is a clear echo in this description of the way in which Husserl, following Brentano, introduced the notion of intentionality. In Chapter 2 the main focus was to clarify the 'objective pole' of intentionality – the intentional object of which we are aware. Here, the focus is on the 'ego pole' of intentionality, the idea being that in any intentional state there is an awareness not only of the object of that state, but of the subject also. In being aware of something I am thereby aware of myself as so aware.

The empirical ego discussed in *Logical Investigations* is an object of awareness and, according to Husserl, is presented in a way analogous to an external thing. This is not so, Husserl tells us, of the pure ego, which

> does not present itself merely from a side [...]. Instead, the pure Ego is given in absolute selfhood and in a unity which does not present itself by way of adumbrations; it can be grasped adequately in the reflexive shift of focus that goes back to it as a center of functioning. As pure Ego it does not harbour any hidden inner richness; it is absolutely simple and it lies there absolutely clear.
>
> (1952, p. 111)

Husserl's position here is essentially Kantian. The pure ego is implicitly given in all intentional experiences. As such, it is given, not as an object (although Husserl claims that it can be made an object), but as the *subject* of the experience in question. That the pure ego is given as identical through all experiences is a necessary condition of their belonging to '*one* stream of mental processes which is mine'. It is given, but not as an intentional object. In short, Husserl accepts the Kantian transcendental ego.

2 Sartre against the transcendental ego

Sartre famously claims that Husserl's change of mind is a mistake, arguing that the transcendental ego is 'superfluous' since consciousness 'unifies itself', and also 'a hindrance' since the postulation of the transcendental ego is 'the death of consciousness' (1937, p. 7). Furthermore, he suggests that it would be an error to suppose that Husserl's later position is truly Kantian. Sartre points out the importance of distinguishing between the claim certainly made by Kant, that each of my experiences must *be capable* of being accompanied by 'I think', and the claim that Sartre denies is made by Kant, that each of my experiences *is in fact* accompanied by 'I think'.

On the first view, to say that the ego is transcendental is simply to express a necessary condition of experience – that, for each of my experiences, I must be able to think of it as mine. We need not countenance a 'transcendental ego', conceived of as something of which I can be (or necessarily am) aware. Furthermore, on this view, it would be a mistake to suppose that such a transcendental ego *explains* the unity of consciousness *via* some form of synthesising activity. Rather, once again, the 'I think' is merely a necessary condition of such unity – where experience is unified then it must be possible for one to think of it as one's own. So, if I see a tree, I must be able to think of this experience as my own. And if I hear a bird singing, I must be able to think of that experience as my own. And if I jointly see a tree and hear a bird singing, then I must be able to think of those experiences as jointly my own. For, as Kant puts it, 'otherwise I would have as multicoloured, diverse a self as I have representations of which I am conscious'.

On the second view, which Sartre takes to be Husserl's, this unity of experience is *explained* by the presence within experience of the transcendental ego. That is, the experiences are unified *because* they are experienced as possessed by a single transcendental ego which is given as identical throughout the stream of experience. This ego is not given as an object of experience for, if it were, that would simply be one more experience in need of unification with the rest. Rather, it is given 'behind each consciousness, as the necessary structure of these consciousnesses, whose rays fall on to each phenomenon that presents itself' (Sartre, 1937, p. 5).

The accuracy or otherwise of Sartre's interpretations of both Kant and Husserl can, for present purposes, be put aside. What is of primary concern is whether Sartre provides us with a compelling case against such a view. We can break this down into two moves. First, he argues that we have no need to appeal to the alleged unifying power of the transcendental ego, since both the unity and individuality of experience can be otherwise accounted for. Second, he claims that both the evidence of reflection and also the correct account of intentionality tell against the view that the ego is implicitly given within all experience as its subject. Rather, as he puts it, we can conceive of, and are in fact continually aware of, 'consciousnesses that are absolutely impersonal' (1937, p. 5). We can look at each of these moves in turn.

2.1 The unity of experience

Sartre suggests that the Husserlian conception of intentional experience is all that is required for an account of the unity of consciousness:

> The unity of the thousand active consciousnesses through which I have added, now add, and will add in the future, two and two to make four is the transcendent object, 'two and two make four' [...]. The object is transcendent to the consciousnesses that grasp it, and it is within the object that their unity is found.

> (1937, p. 6)

Sartre should not be understood here as claiming that all that is required for two experiences to be unified is that they be directed toward the same object. For, obviously enough, two different people can both be aware of the same thing without their experiences thereby being given *together* in the way characteristic of the unity of experience (cf. Priest, 2000, p. 35). Rather, we should remind ourselves of the Husserlian account of the experience of a thing as identical through a changing stream of experience. For example, as I walk around something the way that it appears changes. In Husserl's terminology, it is given through varying adumbrations. Nevertheless, it is given as one and the same thing. That is, the thing appears to be the same.

This is crucial to understanding Sartre's contention that the Husserlian account of intentionality is sufficient to account for the unity of consciousness. If two of my experiences are directed towards the same object and present it *as identical*, then those two experiences are unified, given *together*. Whilst it is true that two different people can both experience the same object, arguably their experiences are not connected by the structure that Husserl attributes to our ordinary perceptual awareness of things as remaining constant through varying appearance.

I say 'arguably' above since it can be objected that precisely such a structure – the awareness of a thing as identical to the thing simultaneously experienced by another – is, in fact, accepted by Husserl and forms the basis of his claim, in *Cartesian Meditations*, that intersubjective experience is more fundamental than experience as of an 'objective' world (this claim will briefly arise again in Chapter 9). As he puts it there, '[t]he experiential phenomenon, Objective Nature, has, besides the primordially constituted stratum, a superimposed second, merely appresented stratum originating from my own experiencing of someone else' (1931, p. 124). This objection is a serious one and calls into question the legitimacy of Sartre's claim that the unity of experience can be simply accounted for from within the Husserlian conception of intentionality. Since, however, its appreciation depends on difficult questions concerning the experience of others, I will not pursue it further here.

A more straightforward objection to this picture is that even if it accounts for the unity of experiences of a single thing, it does nothing to account for the unity of experiences of distinct things. These, of course, are not given *as identical*. As Priest points out, 'if this were not the case one person could not experience different things [as different]' (2000, p. 35). Once more, however, Sartre can appeal to Husserl's account of intentionality to answer this worry. Recall from Chapter 3 that, according to Husserl, things are experienced against a background or 'external horizon'. That is, whenever one experiences a thing, one experiences it *in its relation to* the numerous other objects of which one is aware. Most notably, one will be aware of the spatial relations holding between the things of which one is currently perceptually aware – I am aware of the tree as located between the window and the neighbouring houses. Furthermore, this account can be extended beyond the realm of perceptual experience to incorporate, for example, thought. For, when I close my eyes and consider the tree that I have recently seen, it is implicitly given to me *as* the tree that I have recently seen – as *that very* tree.

The above account incorporates both the unity of experience at a time and the unity of experience over time. It is worth pausing briefly to note that Sartre is offering, admittedly in a highly sketchy way, an account of unity that can be seen as a generalisation of Husserl's account of time-consciousness. This is important since it indicates a way in which Sartre is not just appealing to the unity of the experienced world to account for the unity of experience, but is also presupposing a particular way in which experiences are directed toward one another. As we saw in Chapter 5, Husserl saw every experience as involving a threefold structure of primal impression, retention, and protention. In this way, at each moment experience is directed not only at the present phase of some event but also at past and future phases *and also to experiences of them*. As Sartre puts it:

> It is consciousness that unifies itself, concretely, by an interplay of 'transversal' consciousnesses that are real, concrete retentions of past consciousnesses. In this way, consciousness continually refers back to itself: to speak of 'a consciousness' is to speak of the whole of consciousness.

(1937, p. 7)

There are, of course, different accounts of how it is that experience can be unified over time – of how we can perceive events *as events*. Husserl's account, Sartre tells us, proceeds entirely in terms of intentional experiences being directed towards both a unified world and each other, without recourse to the transcendental ego. Sartre can be seen as simply pointing out that essentially the same resources – a unified world and the self-directed nature of experience – are available to account for unity not only over time, but also at a time. Just as the Husserlian account of time-consciousness situates each perceived event-phase within an experienced temporal context, so the account of horizonal awareness situates each perceived thing in an experienced spatial context.

Sartre's claim that intentional directedness towards a unified world of objects is what unifies experience might, then, seem rather plausible and so leave no work for the transcendental ego to do. There are, however, cases that are tricky to accommodate within this overall account. Consider, for example, the objects that I imagine. If I imagine a golden mountain, one might object, it need not be given as in any way related to the world around me. Or consider Sartre's own example of a mathematical object. Such things, being abstract, are surely not presented as spatially (or temporally) related to the perceptible world around me. They occupy, as it is sometimes put, a different realm.

There are a number of ways in which a Sartrean might respond to these sorts of cases. One option would be to claim that the objects of such experiences are, in fact, given as related to those occupying the surrounding world since they fall under concepts that are presented in ordinary non-imaginative and non-mathematical experience. So, for example, in imagining a golden mountain, it is given to me as related to other golden things in virtue of being made of the same stuff. Similarly, in thinking '2 + 2 = 4', I am aware of the sum as the same as the number of empty coffee cups on my desk, for example.

If one finds this implausible, another option would be to claim that whilst the objects of imagination and mathematical thinking are not given as related to the objects of, say, visual experience, the experiences themselves certainly are given as related to other experiences. Thus, in thinking 2, I am aware of that episode of thought as being preceded by an episode of thinking + and, before that, of an episode of thinking 2. I am also potentially aware of a coming episode of thinking =, and so on. Thus, here we have unity not by way of 'the object' but by way of 'consciousness continually [referring] back to itself'. Obviously, such an account of unity presupposes some form of self-awareness, since it proposes that we are aware of experiences themselves, not just their objects, as unified. We will pursue this proposed relation between awareness and self-awareness in §3.

2.2 The individuality of experience

The question of the individuality of experience is the question of what accounts for the fact that some experiences are mine and others are yours. Since, on Husserl's later view, the pure ego is given as identical through all experiences, it is a necessary condition of their belonging to '*one* stream of mental processes which is mine'. We might then answer the question of individuality by claiming that an experience is mine if it is given to me as mine, as belonging to my pure ego, and yours if it is given to you as yours, as belonging to your pure ego. Sartre, of course, rejects any such answer, claiming that

> the individuality of consciousness evidently stems from the nature of consciousness. Consciousness [...] constitutes a synthetic, individual totality, completely isolated from other totalities of the same kind, and the *I* can, clearly, be merely an *expression* (and not a condition) of this incommunicability and this inwardness.
>
> (1937, p. 7)

It is not immediately clear what Sartre has in mind here but a natural way to read him would be as maintaining that each individual subject of experience is constituted by the totality of experiences that are intentionally directed to each other in the ways outlined in the previous section. My present experience retains only *my* past experience and yours retains only *yours*. Thus, facts about the intentional directedness of experiences to others are all that is required to account for the individuality of consciousness.

There is much that can be said about this suggestion but we should be careful, at this point, not to stray from phenomenological terrain. For the question of what it is that makes some experiences yours and other experiences mine, the question of individuality, concerns how subjects of experience are individuated and is, on the face of it, a metaphysical rather than a purely phenomenological question (at least, on a Husserlian conception of the bounds of phenomenology). As such, even if Hume, Kant, and others have an interest in answering it, it is not clear that such an answer ought to form part of a purely phenomenological investigation.

There is, however, a related question to which a phenomenological investigation surely ought to offer an answer. This is the question of what it is for my experience to be given to me *as mine* and yours to be given to me (and, presumably, to you) *as yours*. We might call this the question of the *given* individuality of experience. As before, we can easily imagine an answer to this question in terms of the transcendental ego: what it is for my experience to be given to me *as mine* is for it to be given as belonging to my pure ego. And, once more, this is an answer that will be rejected by Sartre. Exactly how Sartre will respond to this question is something to which we shall return when looking at his distinction between reflective and pre-reflective consciousness.

2.3 Selfless experience

Sartre claims that the evidence of reflection tells against the view that the ego is implicitly given within each experience as its subject. When we reflect on past experience, Sartre suggests, this is clear:

I was just now absorbed in my reading [...] while I was reading there was a consciousness *of* the book, *of* the heroes of the book, but the *I* did not inhabit this consciousness, it was merely consciousness of the object and non-positional consciousness of itself.

(1937, pp. 11–12)

Again,

When I run after a tram, when I look at the time, when I become absorbed in the contemplation of a portrait, there is no *I*. There is a consciousness of the *tram-needing-to-be-caught*, etc., and a non-positional consciousness of consciousness. In fact, I am then plunged into the world of objects, it is they which constitute the unity of my consciousnesses, which present themselves with values, attractive and repulsive values, but as for *me*, I have disappeared.

(1937, p. 13)

Sartre's claim here is simply that as I go about my ordinary, everyday business, absorbed in the world and the tasks at hand – what Husserl calls the natural attitude – I am aware of no pure ego as the subject of my on-going stream of unified experience. In this respect, at least, Sartre seems to endorse a broadly Humean picture (although see Williford, 2011, for some caution here). If this is right, then it cannot be that awareness of a pure ego is necessary for either the unity or the given individuality of experience.

This, of course, is a claim supposedly supported on phenomenological grounds. Reflection on past experience is supposed to tell us that in our ordinary, unreflective lives we are not aware of a pure ego. But, of course, this is exactly the opposite of Husserl's claim. According to him reflection reveals precisely the opposite. At this point, we might become somewhat pessimistic about the phenomenological method itself. Is this not simply an unproductive clash of intuitions? It is hard to resist the suspicion that the victor in this dispute will be the one who succeeds in thumping the table the hardest.

Such pessimism would, I think, be premature. What is required to make progress here is some explanation, by one side or the other, of how it is that whilst the presence (or otherwise) of a pure ego is evident *via* reflection, it is nevertheless possible for phenomenologists to err on this matter. And, in fact, this is exactly what Sartre offers us, amongst other things, in his account of the distinction between reflective and pre-reflective experience. It is in this account, also, that we shall find both Sartre's answer to the question of given individuality and an explanation of a curious phrase, 'non-positional awareness', occurring in both of the above quotations. Before we look at that distinction, however, we should consider Sartre's claim that it is not simply the evidence of reflection that tells against the transcendental ego, but also the nature of experience itself as purely intentional.

2.4 Transparency and the death of consciousness

Sartre claims that the correct account of intentionality tells against the view that the ego is implicitly given within all experience as its subject. This is the claim, already discussed in Chapter 3, that experience is empty or, to use a term familiar from contemporary discussions, transparent. Husserl claims that perceptual experience has both intentional and non-intentional, sensory features. Sartre denies this, suggesting that the phenomenology of perception need appeal only to intentionality – to what and how worldly objects are presented. On this view, reflection on the character of perceptual experience reveals nothing but the things of which we are aware and their properties – perceptual sensations are a philosopher's invention. As Tye puts the view for the case of vision: 'When you introspect your visual experience, the only particulars of which you are aware are external ones making up the scene before your eyes' (2002, p. 139).

In Chapter 6 we saw Sartre apply this claim to the realm of imaginary experience, claiming that if imagining involved an inner awareness of mental images, then '[c]onsciousness would cease to be transparent to itself; everywhere its unity would be broken by the inadmissible, opaque screens' (1940, p. 6). Sartre makes just this move in the present context of self-awareness, claiming that the pure ego,

> [i]f it existed would violently separate consciousness from itself, it would divide it, slicing through each consciousness like an opaque blade. The transcendental *I* is the death of consciousness [...]. Everything in consciousness is [...] clear and lucid: the object lies opposite it, in its characteristic opacity, but consciousness, for its part, is purely and simply the consciousness of being consciousness of its object: such is the law of its existence.
>
> (1937, p. 8)

According to 'the law of its existence' experience has only intentional features. Reflection reveals an intentional directedness towards objects, but no self that is so directed. Sartre's position here might be fleshed out by following a recent discussion by Howell (2010), who claims that, being transparent, experiences are not 'object presenting' where, roughly speaking, a property is object presenting when an experience of it is an experience of it *as* a property of the object that possesses it (2010, p. 474). Typically, properties are object presenting. One experiences shapes, colours, etc. as properties of their objects. Not so with conscious states:

> from the unreflective first-person perspective, a subject's mental properties do not present themselves as properties of the subject. While he is aware of them in some sense, they are not in fact salient to the subject *as his properties*: they are phenomenologically exhausted in their presentation of the world.
>
> (2010, p. 476)

Since experiences are transparent, not object presenting, in being aware of them one is not thereby presented with oneself. We can thus move from the transparency of perceptual experience to the Sartrean claim that reflection reveals no pure ego.

This picture may be challenged, however. There are at least two difficult cases: bodily sensation and visual experience. Taking bodily sensations first, a plausible assumption is that, in Howell's sense, they present my body. That is, in being aware of a headache I am presented with a particular part of my body. But if, as many philosophers believe, I am identical to my body, then a pain would present a thing which is, in fact, me. Importantly, this is not yet to say that in being aware of a headache as a property of my body I am thereby aware of it *as* possessed by myself (a pure ego), but the fact that a headache is surely a property of me does push in that direction (Brewer, 1995). Questions about the relation between self-awareness and the awareness that each of us has of our own bodies will be addressed in the next chapter. The point here is just that, in the case of the body, transparency does not obviously tell against, and may in fact speak in favour of, the pure ego.

A similar point applies to visual perception. We are aware not only of the shapes, sizes, and colours of visually presented objects, but also of their locations. Indeed, one of the most striking features of the visual scene is that it is egocentric: the objects that I visually experience are presented as occupying locations relative to mine. Some are near, some are far, some are straight ahead, some are to the left, etc. These spatial relations relate experienced objects to the location occupied by me. Some hold that they relate them to the perceiver themselves. For example, Bermúdez claims that '[v]ision presents the world in a fundamentally egocentric and perspectival way. The embodied self appears in visual perception as the origin of the field of view' (Bermúdez, 2011, p. 160). On such a view, the transparency of visual experience to its objects would appear to mandate the inclusion of the self in the content of visual experience, since the self would be presented as one of the things to which experienced objects are spatially related. Once more, we can conclude that transparency considerations do not unambiguously support the Sartrean view.

In response to this, Sartre may insist that we be careful to distinguish between the transcend*ental* and transcend*ent* ego. The suggestion above, regarding both bodily awareness and visual experience, is that the self is presented as an object of experience: bodily awareness is an awareness of my bodily-self as possessing various properties (sensations of pain, etc.); visual experience is an awareness of my bodily-self as standing in various spatial relations to surrounding objects. But this sounds rather unlike Husserl's description of the pure ego. In these descriptions, the self is one of the things one is aware of in bodily and visual experience respectively. But, according to Husserl, the ego is not given in the content of experience, but as that which *has* that experience. In his words, 'I take myself as the pure ego insofar as I take myself purely as that which, in perception, is directed to the perceived' (1952, p. 103).

In short, the self that is here being claimed to be given in bodily and visual experience is not the transcendental ego, but rather the transcendent ego, an entity, 'neither formally nor materially *in* consciousness: it is outside, *in the world*' (Sartre, 1937, p. 1).

Matters are somewhat murky here and are complicated by the fact that the above Sartrean response is presented outside of the context of his distinction between unreflected, reflected and reflecting consciousness. I have been putting this issue to one side for some time. Now it is time to bring it to the fore.

3 Unreflected, reflected, and reflecting experience

In Chapter 1 we saw that, according to Husserl, the phenomenological reduction serves to draw our attention away from the objects experienced towards our experience of them. 'Pure reflection' consists in turning our attention towards our experience itself. In this way we attend to *phenomena*: experiences and their objects *as* experienced. Phenomenology, on Husserl's view, is to be undertaken from within the reflective attitude which can be contrasted with the natural attitude within which our lives otherwise tend to take place.

Reflection, minimally understood, is simply attending to one's own experience and thereby taking it as an explicit theme. Construed in this way, reflection would seem to be indispensable to the phenomenological enterprise. For how else could we describe experience other than by taking it as an explicit theme? Describing it *just is* one way of focusing our attention on it. But, for the most part, our lives are lived out without any such explicit focus of attention on experience. Rather, we are primarily concerned with the world around us, with our various tasks, and so on. This 'pre-reflective' experience is of particular interest to Sartre, who maintains that, in certain respects, including the present context of self-awareness, reflecting on our own experience actually has the consequence of altering it. As he puts it, 'reflection *modifies* spontaneous consciousness' (1937, p. 13). If this is so, then it is surely imperative that we can show that there is a way of describing pre-reflective awareness *as such*. For, if this were not the case, it is unclear how we could ever know anything about it as it exists *un*modified, or indeed that reflection has an effect of this sort.

The distinction between reflective and pre-reflective experience is also crucial for tying together a number of loose ends. In what follows, I set out the essentials of Sartre's distinction between the reflective and the pre-reflective; indicate the role that he thinks reflection plays in our awareness of the transcendent ego; suggest how the reflective/pre-reflective distinction may answer both the question of the given individuality of consciousness and the worry of how phenomenologists could have disagreed over an issue as elementary as whether there is awareness of a pure ego; and, finally, return to the question left hanging at the end of the previous section as to the relation between bodily and visual experience and the transcendent ego.

3.1 Non-positional consciousness of consciousness

In ordinary, everyday experience we are primarily concerned not with experience but with Objects. That is, we do not typically reflect on our experience, explicitly attending to it. This is not to say, however, that we are for the most part entirely oblivious to our own experience. If we were, then we would not be able to offer an account of it. When going about our daily business, whether having a conversation, typing, or brushing our teeth, we are implicitly aware of our own experiences. Whilst our attention is directed towards Objects, we also have a tacit awareness of ourselves as experiencing them. This is what Sartre refers to as 'non-positional consciousness of consciousness', describing the essence of experience as 'consciousness of being consciousness of its object'.

Although the account of self-awareness that Sartre presents in *Being and Nothingness* differs in a number of respects from that offered in *The Transcendence of the Ego* (which is our focus here), §3 of the Introduction of the later book does provide a useful elaboration of this 'consciousness of consciousness':

> [A] necessary and sufficient condition for a knowing consciousness to be knowledge *of* its object, is that it be consciousness of itself as being that knowledge. This is a necessary condition, for if my consciousness were not consciousness of being consciousness of the table, it would then be consciousness of that table without consciousness of being so. In other words, it would be a consciousness ignorant of itself, an unconscious – which is absurd. This is a sufficient condition, for my being conscious of being conscious of that table suffices in fact for me to be conscious of it.
>
> (1943, p. xxviii)

On this view, a certain form of second-order awareness is a necessary (and sufficient) condition of consciousness, or experience. If I am conscious of a tree, I must also be aware of my awareness of a tree. So far, this may seem like contemporary higher-order thought theories of consciousness (Rosenthal, 2005; Gennaro, 2002), according to which a state of mind is conscious if and only if I also have a higher-order thought to the effect that I am in it. To read Sartre in this way, however, would be a mistake. For one thing, Sartre, rightly or wrongly, supposes such a view to result in an infinite regress, since the higher-order thought must itself be made conscious by a yet higher-order thought, and so on (1943, p. xxviii). Secondly, such a higher-order thought would 'posit', or have as an object, the first-order state and, as Sartre puts it, 'pass judgement on the consciousness reflected-on' (1943, p. xxix). But, claims Sartre, non-positional consciousness does not 'pass judgement' on my conscious state – it does not take it to be some way or other – but simply presents it as it is. Non-positional consciousness of consciousness is, in Sartre's view, simply consciousness – exhausted by the fact that my experiences are mine or, equivalently, given *to me*, and not explicable in terms of there being a further experience taking the original experience as its object. Summing this up, Sartre writes:

> We understand now why the first consciousness of consciousness is not positional; it is because it is one with the consciousness of which it is consciousness. At one stroke it determines itself as consciousness of perception and as perception.
>
> (1943, p. xxx)

Such non-positional consciousness is, claims Sartre, selfless. This much is clear from the example of reading mentioned above. When reading, Sartre claims, there is, 'consciousness *of* the book, *of* the heroes of the book', but no ego. There is 'merely consciousness of the object and non-positional consciousness of itself' (1937, pp. 11–12).

We are now in a position to formulate a deflationary, Sartrean answer to the question of the given individuality of consciousness. This, recall, is the question of what it is for my experience to be given to me *as mine*. Those who suppose that we are aware of a pure ego may claim that for my experience to be given to me *as mine* is for it to be given *as belonging to the pure ego*. A Sartrean, on the other hand, might respond by claiming that, since my experiences are non-positional experiences of themselves, they are not nothing to me, I am implicitly aware of them simply by having them. And there is no further question of what it is for them to be given *as mine*. There is simply the question of what it is for me to have an experience which, as we have seen, Sartre answers in terms of my being intentionally directed to worldly objects. The given individuality of consciousness is nothing more than the fact that some experiences are mine and others are yours, and it gives no support to defenders of the transcendental ego.

Some of the above reasoning may be disputed by appeal to the work of Zahavi (2005, Ch. 5), who defends the idea that the pre-reflective consciousness of consciousness in fact involves a form of 'minimal selfhood' that, ironically, tells against this Sartrean position. He writes that

> In order to have a self-experience, it is [...] not necessary to apprehend a special self-object, it is not necessary to have a special experience of self alongside yet different from other experiences; rather what is required is simply an episode of pre-reflective self-awareness. What is needed is an acquaintance with the experience in its first-personal mode of presentation.
>
> (2005, p. 126)

Zahavi's claim here is that whenever I am conscious of one of my experiences by way of simply having it – in his words, 'in its first-personal mode of presentation' – then the experience is implicitly given *as mine*. This 'sense of mineness', which amounts to an awareness of a 'minimal self', is 'not something attended to; it simply figures as a subtle background presence' (2005, p. 126). In this way, the Husserlian claim that through all experience we are aware of a 'pure' or 'minimal' self is vindicated. Despite his intentions, then, Sartre's endorsement of pre-reflective consciousness of consciousness in fact amounts to an endorsement of exactly the sort of self-awareness that he explicitly rejects. The minimal self is given in all experience, without being given as an object.

If this is correct, then the position sketched by Sartre in *The Transcendence of the Ego* is unstable. But one might well question Zahavi's line of reasoning. The crucial move is from the idea that Sartre's 'consciousness of consciousness' is 'one with the consciousness of which it is consciousness', to the claim that it is a form of *self*-awareness, an awareness of a minimal self. On the face of it, Sartre's claim is that each experience is, implicitly, also an experience of *it*self. In Zahavi's hands, however, this becomes the claim that each experience is, implicitly, an experience of the minimal self, of *one*self. But why, we may ask, is it not sufficient to claim that one's awareness, from within, of an experience just amounts to one's having it? Why, in addition, must one be aware of it *as one's own*? Of course, one does ordinarily believe one's experiences to be one's own, but it is not entirely obvious that this is built into the very structure of experience. More needs to be said, it seems, before we can resolve this issue.

3.2 Reflecting and reflected consciousness

Non-positional consciousness of consciousness is, in the usual run of our daily lives, not reflected upon. It is, rather, unreflected. When we reflect, however, we direct our attention toward our experience, taking it as an explicit theme. In such cases we can distinguish between the reflecting and the reflected consciousness. Reflecting consciousness is a second-order consciousness directed to first-order experience. Reflected consciousness is first-order experience *as it is presented* to reflecting consciousness. In the previous sections I outlined Sartre's account of selfless unreflected consciousness. Sartre's account of reflecting consciousness which – as with unreflected consciousness – is accompanied by an implicit consciousness of consciousness, follows suit. Reflecting consciousness, he indicates, is itself selfless, since 'all reflecting consciousness is itself unreflected' (1937, pp. 10–11).

Sartre tells a very different story, however, with regard to reflected consciousness, asserting that 'no one dreams of denying that the *I* appears in a reflected consciousness' (1937, p. 12). This is the 'transcendent ego' – equivalent to the Husserlian 'empirical ego' – of Sartre's title. When reflecting, I am aware of the transcendent ego that 'affirms its permanence beyond that consciousness' (1937, p. 14). Just as Objects are given *via* adumbrations as transcending my awareness of them, so the ego is given *through* experiences as transcending them, presenting 'itself as an opaque reality whose content would need to be unfolded' (1937, p. 15).

A useful analogy can be drawn here between the role of reflection and a certain use of cameras. A film represents the world from a particular location, the location of the camera. But the camera itself is not represented (much like, as Wittgenstein famously pointed out, the eye is not represented in the visual field). This is akin to unreflective consciousness, which Sartre claims is selfless. But now consider a second camera – like reflective consciousness – panning back to reveal the original camera. We now have a representation of the camera but not as at a location from which the world is represented but as at a location within the represented world. According to Sartre's account, reflection operates on something like this model.

Unreflected experience, then, of which we have an implicit consciousness, is selfless. Yet, 'the *I* [...] gives itself *through* reflected consciousness' (1937, p. 15). Since, however, reflected consciousness just is unreflected consciousness as it appears to reflection, it must be that reflection on my own experience actually modifies it, *generating* the transcendent ego. This is exactly what Sartre claims, rhetorically asking whether

> it might not be precisely the reflective act that brings the *me* into being in reflected consciousness? This would explain how all thinking grasped by intuition possesses an *I*, without running into the difficulties [that beset Husserl's view].
>
> (1937, p. 11)

The transcendent ego is, according to Sartre, a product of reflection. Yet, and here is the twist that allows Sartre to offer an explanation of how it is that Husserl could have got it so wrong, 'it manifests itself as the source of consciousness [...] and by this fact, too, it is immediately deceptive' (1937, p. 15). That is, whilst we can be certain, *via* a consideration of unreflected consciousness, that the transcendent ego is the product of reflection, things *appear* quite the reverse, with the reflection seemingly the product of the ego. Thus, Sartre offers an explanation of how it is that phenomenologists have been misled into supposing that there is a pure, transcendental ego inhabiting all experience.

Is Sartre's account of unreflected, reflected, and reflecting experience acceptable? I will mention three ways in which the account can be challenged. The first draws on Heidegger's remarks on self-awareness in *The Basic Problems of Phenomenology*, in which he endorses a form of unreflective self-awareness:

> Dasein, as existing, is there for itself, even when the ego does not expressly direct itself to itself in the manner of its own peculiar turning around and turning back [...]. The self is there for the Dasein itself without reflection and without inner perception, *before* all reflection.
>
> (1927b, p. 159)

How, according to Heidegger, is the self 'disclosed' in unreflective experience? It is a matter, he tells us, of our being presented with ourselves from the world of things. As Heidegger puts it: '[Dasein] finds *itself* primarily and constantly *in things* because, tending them, distressed by them, it always in some way or other rests in things. Each one of us is what he pursues and cares for' (1927b, p. 159). Recall that, on Heidegger's view, the world is a meaningful context in which things can be encountered. That context, and those things, are given first and foremost in terms of how they are significant *for us*. Thus, one is not in the world in the way that a table is inside a house, rather one *dwells* in the world. Our being is being-in-the-world and that world is thereby given to me as *mine*, structured in terms of my goals and cares. Since such goals and cares at least partly define *who I am* ('each one of us is what he pursues and cares for'), in being unreflectively aware of the world in which I dwell, I am thereby aware of myself. Whilst this form of what we might call 'environmental self-awareness' is very different from the variety of self-awareness that Husserl endorses, it nevertheless looks to be inconsistent with Sartre's insistence on the selflessness of unreflective experience, and so demands a response from the committed Sartrean.

A different worry concerns the discussion in §2.4 of the relation between bodily and visual experience, on the one hand, and the transcendent ego, on the other. Recall that some philosophers have suggested that bodily and visual experience amount to a form of self-awareness of the 'bodily self'. The thought that this might pose a problem for the argument from transparency to selflessness was countered, on Sartre's behalf, by suggesting that this bodily self is to be identified with the transcendent, rather than the transcendental, ego. Thus, the view that bodily and visual experience are forms of self-awareness can be made consistent with the Sartrean position. However, now that we have Sartre's distinction between unreflected and reflected experience in view, this move may appear too quick. Both bodily and visual experience, after all, may occur unreflected. As such, the Sartrean is committed to denying that unreflected bodily and visual experience are forms of awareness of the bodily self. But it is not obvious what would motivate such a claim. The reasons for supposing that the two forms of experience are forms of self-awareness – the role of the body as the location of sensation and the perspectival nature of vision, respectively – surely apply regardless of whether or not the experience is reflected upon. The Sartrean does, therefore, have some work to do to defuse this worry.

I end with one more worry, raised by Gardner (2009, pp. 14–15), that can be seen as the redirection of one of Sartre's criticisms of the Husserlian view back towards his own account. Sartre points out that 'if the *I* is part of consciousness, there will then be *two I*'s: the *I* of reflective consciousness and the *I* of reflected consciousness' (1937, p. 15). That is, the Husserlian accepts both the transcendental and empirical ego. But this creates the question of how these may be related. Whilst Sartre's contention that this question presents a problem that is 'quite simply insoluble' may be an overstatement, there does seem to be a genuine issue here and, in fact, one that arises, in a slightly different form, for Sartre's own view. Sartre claims that reflecting consciousness is selfless – 'impersonal' – and that the transcendent (and only) ego is manifested through reflected consciousness. This transcendence is a matter of the ego's being an object for consciousness, such that Sartre can say that 'the Ego [...] is a being in the world, like the Ego of another' (1937, p. 1). But, of course, my ego and its experiences are not given to me in *exactly* the same way as are those of the other. My ego is presented to me, precisely, as *me*. One way to bring this out would be to say that, in reflection, I am aware that the reflected and reflecting consciousness *belong to the same subject*, i.e. myself:

> If reflection did not grasp itself as 'of the same subject' as the consciousness reflected on, then the mental stream which is presented to me by reflection on my consciousness would have the same indifferent, alien character as the external world.
>
> (Gardner, 2009, p. 15)

Sartre himself comes close to suggesting as much in his claim that the transcendent ego 'manifests itself as the source of consciousness' (1937, p. 15) since, presumably, if the ego is given as the source of consciousness – both reflected and reflecting – then consciousness must appear as possessed by the ego. But, if this is right, then surely there must be some level of self-consciousness at the reflecting level? For how could reflecting consciousness be, even implicitly, given as belonging to the same subject as reflected consciousness if it were entirely impersonal, i.e. not given as belonging to a subject at all? This is a question that requires an answer as, without one, it may be that the Sartrean is pushed back to something like Zahavi's account of 'minimal selfhood' after all.

4 Conclusion

The experience of oneself and one's own experiences is a fundamental element of Husserl's conception of the phenomenological method. Even if, with both Heidegger and Sartre, one denies that all phenomenology is reflective, self-awareness must surely be seen as central to our experiential life. But it is less clear what it is to be aware of oneself. Husserl's broadly Kantian account clashes with Sartre's broadly Humean account in a fundamental way and contemporary writers continue to disagree. Self-awareness and its close relative the unity of conscious experience are subjects never far from the surface in all phenomenological investigations.

8 Experiencing embodiment

> A living body only perceived outwardly would always be only a particularly
> disposed, actually unique, physical body, but never 'my living body'
>
> Stein, *On the Problem of Empathy*

As I sit typing at my desk I feel the keys under my finger tips, see my hands
moving before me, feel the pressure of the chair beneath me and of the clothes
that I am wearing. I am aware that my legs are crossed, despite the fact that
they are hidden from view. Sitting up straight, I feel my chest move as I inhale.
In short, I am aware of more than just things in the surrounding world, I am
aware of my own body. Indeed, I am aware of it in a particularly intimate way
in which I am aware of no other thing. As we shall see, the discussion in
previous chapters implicates our experience of embodiment in ways that have
yet to be fully spelled out. In this chapter we will look more closely at the
experience of embodiment, focusing on the awareness that each of us has of
our bodies 'from the inside'; the sense that each of us has of one's body being
one's own; the relation between bodily awareness and self-awareness; and the
way in which embodiment informs our experience of the surrounding world.

I *Körper* and *Leib*

Descartes famously noted that: '[n]ature also teaches me, by these sensations
of pain, hunger, thirst and so on, that I am not merely present in my body as
a sailor is present in a ship, but that I am very closely joined and, as it were,
intermingled with it' (1641, p. 116). His point is the phenomenological one
that the way in which I experience my own body differs from the way in
which I experience other things. This is not to say that I *cannot* experience my
body as a 'mere thing', but rather that there is also, and perhaps more fundamentally, an experience of one's own body that is profoundly different from
that sort of awareness.

In this vein, Husserl distinguishes between the body given as a thing
(*Körper*) and the body as it is *lived* (*Leib*). This is brought out in Husserl's
description of the experience of touching a part of one's own body. Using my

right hand to touch my left, I become aware of the left hand's shape, solidity, etc. But I also experience sensations in, or on, my right hand itself. Thus:

> the body (*Leib*) is originally constituted in a double way: first, it is a physical thing, *matter*; it has its extension in which are included its real properties, its colour, smoothness, hardness, warmth, and whatever other material properties of that kind there are. Secondly, I find on it, and I *sense* 'on' it and 'in' it: warmth of the back of the hand, coldness in the feet, sensations of touch in the fingertips.
>
> (Husserl, 1952, p. 153)

The first way here is that by which my body is given as a thing, the second as a lived body. My own body as presented in 'outer perception' – in particular, sight and touch – is given to me as a thing: 'the appearances have just the same nexus as do other appearances of things' (1952, p. 152). It is, for example, possible to confuse it with the body of another – relying on outer perception alone, I can in principle take someone else's hand for my own and *vice versa*.

Even through outer perception, however, the body is given as a thing unlike any other. As Stein puts it: 'if we suppose it to be given to us in this manner alone, we have the strangest object' (1917, p. 41). Visually, it is incomplete, exhibiting 'striking gaps' (1917). The invisible parts, for example the majority of my head and back, can of course be touched. But the usual unity of touch and vision – whereby that which can be touched can also be seen – is broken in the case of one's own body. My body is the only object that I cannot approach, and from which I cannot withdraw:

> [a]s long as I have my eyes open at all, it is continually there with a steadfast obtrusiveness, always having the same tangible nearness as no other object has. It is always 'here' while other objects are always 'there'.
>
> (1917, p. 42)

Even with eyes closed and no touching body parts, however, I am still aware of my body, which 'stands there inescapably in full embodiment' (1917, p. 42). Without looking I know whether or not my legs are crossed, whether or not my arms are flailing around, whether or not I am about to fall over. I also have a rough sense of the size and shape of my body. Clearly, in addition to the commonly recognised five senses, we are aware of our bodies 'from the inside'. This is bodily awareness. Whilst the 'outer' senses of vision and touch provide me with an awareness of my body as a thing, we also possess an awareness of the position, orientation, movement and size of our limbs (proprioception and kinaesthesia) and a sense of balance. Through these forms of bodily awareness I experience my body in a way in which I experience nothing else. I experience it as something that I *live*. In Merleau-Ponty's words, my body experienced from without is 'an intersecting of bones, muscles, and flesh compressed into a point of space', whilst from within 'it shoots across space like a rocket' (Merleau-Ponty, 1945, p. 94, translation amended). And, unlike with outer perception, *via* bodily awareness it seems impossible to confuse one's own body with the body of another. On the face of it, if I am aware of a body through bodily awareness, then it is manifestly my own body of which I am aware (cf. Evans, 1982, Ch. 7).

2 The sense of ownership

Not only does each of us have a body but, at least in ordinary circumstances, each of us believes of a certain body that it is our own. More than that, however, each of us is aware of a certain body *as our own*. My body *seems to be mine*. What is involved in this form of awareness of our own bodies? Intuitively, there is a connection between a body seeming to be one's own and one's being aware of it 'from the inside', as a *lived body*. According to Stein, indeed, it is only *via* such inner bodily awareness that such 'affiliation' can be accounted for. As she puts it: '[p]recisely this affiliation, this belonging to me, could never be constituted in outer perception. A living body only perceived outwardly would always be only a particularly disposed, actually unique, physical body, but never 'my living body' (1917, p. 42; cf. Husserl, 1952, §37).

What Stein here refers to as 'affiliation' has more recently become known as the 'sense of ownership' or the 'sense of mineness' (cf. Zahavi, 2005; Bermúdez, 2011). This terminology makes explicit the thought that characteristic of the *lived body* is that it is given *as mine*. In these terms, Stein can be understood as claiming that my body if given to me only *via* sight and touch would not be given to me as *mine*. Whilst Stein doesn't really argue for this, we might rest such an argument on a point sketched in the previous section: only through bodily awareness are we aware of our body in such a way that we cannot mistake something else for it. Through vision or touch I can in principle mistake someone else for myself. But this is not so through bodily awareness. This is the claim that bodily awareness, but not the visual or tactile perception of one's own body, gives rise to judgements that are immune to error through misidentification relative to the first-person pronoun (Evans, 1982, Ch. 7). The question 'Are those legs I see/touch mine?' makes perfect sense, whilst the question 'Are these legs I feel (from the inside) mine?' does not. One reason for this difference might well be that, unlike the senses of vision and touch, the awareness that I have of my body from the inside presents it *as mine*. If my *lived body* is given to me *as mine* it presumably makes no sense to wonder whether or not it is my body of which I am so aware.

Stein does go on to claim that there is, in fact, a 'fusion' between the *lived body* and the seen body, such that the two are identified in experience and one's body comes to be seen *as my lived body* (1917, pp. 44–5). We shall return to this in §3. In the present context, however, we can simply note that such a sense of ownership attaching to the *visually* experienced body would be derivative. Fundamentally, the sense of ownership is associated with the awareness of one's body from the inside, with bodily awareness. This is our current topic and, regarding it, the central question is exactly what the sense of ownership involves. Stein offers a number of features of the lived body that contribute to the sense of ownership. On her view, my body is experienced as mine because: (1) I feel sensations as located in or on it; (2) it defines the outer 'here'; and (3) it is both the direct object of my will and that in which my feelings are expressed. In this section, I briefly consider each of these features. In Stein's own discussion it is unclear what she takes the status of these claims to be. It is tempting to suppose that they are intended to be individually necessary (and perhaps jointly sufficient) conditions. Whether or not this is in fact Stein's view, the plausibility of each condition as either necessary or sufficient is evidently worthy of consideration.

2.1 The lived body as location of sensation

The lived body is *the location of bodily sensation*. Husserl, in his account of the lived body, gives pride of place to this feature, as can be seen from the long quotation in §1 of this chapter (cf. Carman, 1999). Of particular significance is the fact that bodily sensations are experienced 'in a double way'. When I touch something, I have a tactile experience of the properties of that thing but I also experience my hand itself from the inside. Indeed, as Husserl describes it, the very same sensations that serve to present the properties of the touched thing also present my body:

> In the case of the hand lying on the table, the same sensation of pressure is apprehended at one time as perception of the table's surface [...] and at another time produces, with a 'different direction of attention,' [...] sensations of digital pressure. In the same way are related the coldness of the surface of the thing and the sensation of cold in the finger.
>
> (1952, p. 154)

Such experiences are what Armstrong (1962, Ch. 1) calls 'transitive'. They are sensations of something beyond the body part in which the sensation is felt – in this case the properties of the table. They can be distinguished from 'intransitive sensations', such as pains and itches, which seem not to be sensations of anything beyond the body part in which they occur. Whilst my pressure sensation is a sensation of the hardness of the table, my pain does not seem to point beyond the body in this way. Indeed, a long tradition has held that intransitive bodily sensations, such as pain, are non-intentional, purely subjective qualities (Jackson, 1977, Ch. 3).

Of course, a defender of the view that all experiences are intentional will wish to reject such a picture. The most obvious view with which to replace it is that according to which bodily sensations involve perception of the body (Armstrong, 1962; Tye, 2003, Ch. 2). If we think, in this way, of bodily sensations as involving the perception of a body part, then we can think of the lived body as the location of bodily sensation. It seems plausible to think that insofar as I feel a sensation to be located somewhere, I feel my body to extend to that point (Martin, 1995). As Stein says:

> sensation is always spatially localized 'somewhere' [...]. And this 'somewhere' is not an empty point in space, but something filling up space. All these entities from which my sensations arise are amalgamated into a unity, the unity of my living body, and they are themselves places in the living body.
>
> (1917, p. 42)

This singles out the lived body from all other things. Only on the lived body do I experience sensations. As such, we have some reason to suppose that it is bodily sensation that partly grounds the experience of one particular object – my body – as mine.

As a challenge to such an account, de Vignemont (2007) claims that bodily sensation is not necessary for the sense of ownership on the grounds that anaesthesia, which involves the lack of sensation, does not always remove the sense of one's body as one's own. That is, one can possess a sense of one's body part as one's own, even if one feels no sensation in it. Further, some people suffering from asomatognosia – a condition in which patients disown an 'alien' limb – appear to experience sensations in a limb over which they seemingly lack a sense of ownership. If this is correct, then bodily sensations seem not to be sufficient for the sense of ownership. A similar, and perhaps more familiar, challenge to the necessity claim concerns our use of tools. When, for example, I write with a pencil there is some plausibility to the thought that I experience a sensation as located at the tip of the pencil. Try it. But, it might be claimed, the pencil is not thereby experienced as a part of my body. Thus, again, sensations are not sufficient for ownership. It is clear that a proponent of an account of the sense of ownership that places significant weight on sensation must provide some response to these challenges.

2.2 The spatiality of the lived body

Bodily sensations are presented as located in or on my lived body, and thus as *in space*. According to Stein, however, the spatiality of the body cannot be understood on the model of the spatiality of other things. This is grounded in the fact that perception is always *from somewhere*; we experience the world from a spatial location. In short, the perception of space is perspectival. As I look at the tree beyond my window, it is presented as occupying a particular location. The location that it appears to occupy is not specified in terms of absolute coordinates, but rather in relative terms. More specifically, it is given as occupying some location relative to the apparent location of my body. Things appear as to the left, to the right, or straight ahead *relative to my body*. As Husserl expresses the point:

> all things of the surrounding world possess an orientation to the Body [...]. The 'far' is far from me, from my Body; the 'to the right' refers back to the right side of my Body, e.g., to my right hand [...]. I have all things over and against me; they are all 'there' – with the exception of one and only one, namely the Body, which is always 'here'.
>
> (1952, p. 166)

One's own body is thus singled out as the entity located at 'the zero point' of perceptual orientation. But, of course, one's body is not located at a single point in space, it is itself extended throughout a region of space. The perceptual 'here', then, is itself extended. Consequently, the spatial relations between one's various body parts (arms, legs, head, etc.) cannot be given in just the same way as the spatial relations holding between things in one's surroundings. Whilst an outer thing's being to the left is for it to be to the left of my body, my left hand's being to the left cannot be. My left hand *is* my body. In Stein's words,'the living body as a whole is the zero point of orientation with all physical bodies outside it. 'Body space' and 'outer space' are completely different from each other' (1917, p. 43).

One's body parts are given as located in various parts of 'here' with one's bodily sensations given as located in or on the body. In fact, Stein goes beyond this, claiming that one's own body parts themselves are presented as occupying 'various distances from me. Thus my torso is nearer to me than my extremities' (1917, p. 42). Outer space is thus distinguished from bodily space since locations in the former are relative to my *body*, whilst locations in the latter are relative to *me*.

One's own body, then, is presented both as occupying space and – since space is given in body-relative terms – as that in virtue of which we experience space at all. Embodied experience is, on this picture, implicated in the very notion of outer (spatial) perceptual experience. We will return to the way in which the experience of one's body shapes outer perception in §3.

This special relationship with spatial perception singles one's own body out from all other objects of which one is aware. Nothing other than one's lived body is the centre of orientation. But, more than this, there is reason to suppose that the body's role as centre of perceptual orientation is closely related to the sense of ownership. For, on the face of it, it is inadequate to the phenomenology of perceptual experience to say simply that things are given as located relative to *a* body. It seems, rather, more accurate to say that they are given as located relative to *my body*. A ball is thrown towards *me*, slightly to *my* right. I reach *my* right hand out to grasp it. The centre of perceptual space is not given as *a* body but as *my* body. As we saw in the previous chapter, Bermúdez puts this point by saying that '[t]he embodied self appears in visual perception as the origin of the field of view' (2011, p. 160).

If the above is to be believed, then one might suppose that a body's being located at the zero-point of perceptual orientation is both necessary and sufficient for its being given as *my body*. For, if things are experienced as to *my* left or right, then surely the body at the centre must similarly be perceived as *mine*. This thought might be challenged, however, by a consideration of perspectival orientation in pictures and films. When we watch a film, the action is presented from a certain viewpoint (in fact, the location of the camera). But it is far from obvious that such perceptual orientation need involve the presentation of a body – *my body* – as its origin. If this is right, then spatial perception is not sufficient for the sense of ownership. Nor, one might argue, is it necessary. Of course, both visual and auditory perceptual experience have egocentric spatial content with one's own body experienced as located at the origin. But people who lack both sight and auditory perception pose a challenge to the necessity claim, unless there is reason to suppose that they lack the sense of ownership over their own bodies. Once more, if we wish to account for the sense of ownership in terms of the position of the lived body at the centre of perceptual space, we will need to address these issues.

2.3 The lived body as agent

I can move my pen but, to do so, must move my hand or some other part of my body. I can move my hand, however, without having to move anything else. I simply move it and, if asked *how* I do so, I am typically at a loss to provide an informative answer. The lived body is that thing that I may move at will, without thereby having to move some other thing. It is that through which one actively engages in the world. This, once more, singles out the lived body from all other things. My body is given as *mine*, it might be claimed, because it is given as that through which *I* act.

It is a notable feature of such bodily agency, however, that one need not consciously exercise one's will with respect to each and every movement that is responsive in this way to one's will. In Stein's example:

> I can decide to climb a mountain and carry out my decision. It seems that the action is called forth entirely by the will and is fulfilling the will. But the action as a whole is willed, not each step. I will to climb the mountain. What is 'necessary' for this takes care 'of itself'.
>
> (1917, p. 55)

My lived body is immediately responsive to my will but, once set in motion, there is a sense in which it 'knows what to do'. This is a phenomenon that Merleau-Ponty places at the centre of his account of our experience of embodied agency, arguing that:

we must acknowledge, between movement as a third-person process and thought as a representation of movement, an anticipation or a grasp of the result assured by the body itself as a motor power, a 'motor project', or a 'motor intentionality' without which the instructions would remain empty.

(1945, p. 113)

Influenced by Heidegger's account of our everyday comportment as an engagement with *environmental things* or *equipment*, Merleau-Ponty describes such motor intentionality as an engagement towards worldly entities that are *for* something, what he calls 'an ensemble of *manipulanda*' (1945, p. 107). This meaningful 'milieu', ultimately comprising our phenomenal 'world', forms a background against which one's bodily action can take place and be given a motor meaning. My body 'knows' how to carry out the instruction of my will as it itself embodies a comprehension of the significance of the worldly entities with which I am engaged. It is this phenomenon that Dreyfus (1993) calls 'coping'.

The relationship between bodily movement and the will, enabled according to Merleau-Ponty by the body's 'motor intentionality', provides a clear example of the sense in which one's body seems to be closely connected, within experience, to one's psychological life. But it is not the only one. As Stein points out, some bodily behaviour is *expressive* of our emotional life. We experience psychological states as 'poured into' and 'terminating' in expressive behaviour. My blush expresses my embarrassment, my clenched fist expresses my anger, my furrowed brow expresses my worry, and so on. This expressive relation, according to Stein, is given in bodily experience:

> The smile in which my pleasure is experientially externalised is at the same time given to me as a stretching of my lips [...]. Should I then turn my attention to the perceived change of my living body, I see it as effected through a feeling.
>
> (1917, p. 53)

We can distinguish between the experience of a mere bodily movement, as when someone moves my arm, and a willed action. But, according to Stein, we must also recognise a range of cases that sit between these two. In smiling at a joke, I do not experience the movement of my lips as I would were they being manipulated by an outside force, nor do I experience them as moving in accordance with my will. Rather, I experience their movement as the natural termination of my amusement. As with the experience of the will's engagement with the body, this sense of my body as expressive of my mental life singles it out as *mine*. It is *my body*, in part, for the reason that in and on it I experience 'expressive phenomena [that] appear as the outpouring of feeling' (1917, p. 54), *my* feeling.

It seems plausible to suppose that there is a close connection between the fact, discussed above, that spatial perception is perspectival, centred on the lived body, and the fact that one's body is that which I am able to move at will. One way of making this connection is to suppose that to experience something as, say, to the right is to experience it as in *this direction* (imagine those words accompanied by a pointing right hand) (cf. Noë, 2004, p. 87). According to Husserl, there is also a connection between the role of the body in action and the fact that it is the location of sensation. He writes:

> The distinctive feature of the Body as a field of localization [of sensation] is the presupposition [...] for the fact that it [...] is an *organ of the will*, the *one and only Object* which, for the will of my pure Ego, is *moveable immediately and spontaneously* and is a means for producing a mediate spontaneous movement in other things.
>
> (1952, p. 159)

It is reasonably easy to see why Husserl might suppose that the body's role as the location of sensation should be necessary for one to enjoy the sort of immediate control over it with which we are familiar. For it is intuitively plausible to suppose that in order to move my hand, I must have some sense of *where it is*. This sense is precisely an awareness of it from the inside as occupying a certain spatial location. There is, however, some reason to suppose that this connection is not as tight as Husserl may have supposed. As much is suggested by the well-documented case of Ian Waterman (Cole, 1995), who at the age of nineteen lost all sense of the position and movement of his limbs. At first unable to move his limbs, Waterman eventually learned to control his movements visually.

There is a great deal to be said about this incredible case but, on the face of it, the first thing that we can conclude is that even if in order to move a body part at will one must, in some sense, know where it is, and even if that knowledge is usually available to us from the inside *via* bodily awareness, it is in principle possible for outer perception of one's body to play that role. Thus, Husserl's claim above may seem implausible.

The case of Ian Waterman does not tell, however, against the claim that the sense of experienced agency is either necessary or sufficient for the sense of ownership over one's own body. To argue against the necessity claim, one would need to consider examples in which a person lacked agency – for example, the complete loss of motor function – but retained the sense of ownership, perhaps because they retained the capacity to experience bodily sensation in their paralysed body parts. To challenge the sufficiency claim, on the other hand, one would need to consider cases in which a person retains the capacity for agency but without the sense of ownership. At the more speculative end of things, we might conceive of cases of telekinesis, in which non-bodily objects are moved through sheer force of will, but are not thereby experienced as parts of one's body. Closer to reality, we might consider cases of alien hand (see, for example, Moro *et al.*, 2004) where, seemingly, a body part is experienced 'from within' but not *as one's own* and, at least in some cases, can still be moved at will. Once more, if bodily agency is to be the crucial factor in the sense of ownership, then these are challenges that will need to be addressed.

3 Embodied subjectivity

Embodiment represents something of a puzzle for those wishing to divide the contents of the world into subjects and objects. For, on the one hand, one's body seems to be a thing much like any other but, on the other hand, it is – especially in its role as the location of sensation – intimately tied to the subject of experience, *me*. It is unsurprising that so many philosophers reach for terms such as 'the bodily self' in order to describe the way in which our bodies seem at once subjective and objective.

3.1 Is the lived body experienced as a thing?

One is aware of one's lived body in a way different from the way in which one is aware of other things – from the inside. But is the lived body experienced as a worldly thing at all? As discussed in previous chapters, things are given in profile and can be explored. They offer more than is currently seen. One's body itself, however, cannot be explored in just the same way.

> I observe external objects with my body, I handle them, inspect them, and walk around them. But when it comes to my body, I never observe it itself. I would need a second body to be able to do so, which would itself be unobservable.
>
> (Merleau-Ponty, 1945, p. 93)

According to Merleau-Ponty, then, the body is not given as a thing. It is 'the unperceived term at the centre of the world toward which every object turns its face' (1945, p. 84). To appreciate this, it is helpful to turn back to earlier discussions of the ways in which the active lived body is implicated in our awareness of the surrounding world.

First, as we saw in Chapter 3, Husserl claims that I experience objects as having sides that I would see if I *walked* around them, if I *turned them over*, and so on. Walking and turning things over are activities that one accomplishes with one's body. It seems, then, that the awareness of an entity as a thing refers back to one's capacity for bodily activity. Second, as we saw in Chapter 4, Merleau-Ponty's own account of experiencing properties (perceptual constancy) is explicitly couched in terms of one's own bodily agency – to experience a thing as possessing a property is to experience it as inviting certain bodily activities in order to bring it into full view.

If these accounts are along the right lines, the way in which we experience the surrounding world presupposes an awareness of the lived body as our way of navigating that world. But if the lived body is, in this way, a necessary condition of the experience of a world of things and their properties we can see why Merleau-Ponty might deny that one's lived body is presented as a thing. The awareness of one's own body is that in virtue of which there can be things for me, it is that in virtue of which they are 'observable':

> Insofar as it sees or touches the world, my body can neither be seen nor touched. What prevents it from ever being an object or from ever being 'completely constituted' is that my body is that by which there are objects.
>
> (1945, p. 94)

Thus, any awareness of my own body in which it is given as a thing must itself presuppose a prior awareness of it in which it is not given as a thing but in which it is 'our means of communicating with [the world]' (1945, p. 94).

It seems clear, though, that this line of thought depends on a particular view of what it is for something to be given as a thing. On this view, for something to be given as a thing is for it to be given through a variety of perspectives, and 'the always new perspectives are not, for the object, [...] a contingent manner of appearing to us. It is only an object in front of us because it is observable' (1945, p. 92) in this particular way. Further, 'the object is only an object if it can be moved away and ultimately disappear from my visual field' (1945, p. 92). One's lived body, it would seem, fails to meet these conditions. One's lived body is always presented from the same 'perspective', i.e. *from the inside*, and it is not something from which we can take our leave.

One might, however, note that the awareness of one's own body from the inside does seem to present it as the bearer of the primary qualities of solidity, shape, location, or movement. If one were to defend the view that the awareness of something as possessing such primary properties is sufficient for an awareness of it as a thing, then one might be in a position to claim that, despite the fact that the lived body is not observable in Merleau-Ponty's sense, it is nonetheless given as a thing (cf. the list of 'object properties' in Bermúdez, 1998, pp. 72–3).

Touch, for example, gives us an awareness not only of the solidity and location of the touched thing, but also of one's body's own solidity and location. When my arm bumps up against something, I gain an awareness of where my arm is and, were my arm not also given as solid, I would surely not experience its movement as impeded. Further, proprioception gives us an awareness of the body's shape, and kinaesthesia an awareness of the body's movement. As I run, I have a sense of the length of my limbs and the trajectory they trace through space. All of this, it would seem, is *from the inside*. In this way, one might attempt to justify the claim that through bodily awareness one's body does in fact appear as a (shaped, located, solid, etc.) thing. Merleau-Ponty's claim, then, invites further reflection on what is strictly speaking necessary for an awareness of an entity as a thing, thereby deepening the discussion of that question in Chapter 3.

3.2 Bodily awareness and self-awareness

Whether or not the lived body is given as a thing we can also ask whether it is given as a subject. In Chapter 7 we considered the issue of self-awareness. Is there, we asked, an inner awareness of the self? One challenge to Sartre's account of these matters was that bodily awareness provides us with just such a form of experience. It is a (pre-reflective) awareness of my body not just as *mine* but as *me*. One way in which such a case can be made begins from the claim that the body is the location of sensation. As Merleau-Ponty puts it:

> If I say that my foot hurts [...] I mean that the pain indicates its place, that it is constitutive of a 'pain-space.' 'I have a pain in my foot' does not signify that 'I think that my foot is the cause of this pain,' but rather, 'the pain comes from my foot,' or again, 'my foot hurts.'
>
> (1945, p. 96)

On this view, my body is presented as where pains are. Brewer (1995) argues in this way that reflection on the spatial location of bodily sensation shows that bodily awareness is a form of self-awareness:

> In bodily awareness, one is aware of determinately spatially located properties of the body that are also necessarily properties of the basic subject of that very awareness [...] a psychological property of oneself is physically located in or on the body, as a property of the body. Therefore [...] the animal body *is* the conscious mental subject of bodily awareness.
>
> (1995, p. 300)

This, of course, connects directly with the discussion of self-awareness in the previous chapter. If bodily sensations are located properties of one's body and, as Brewer claims, bodily sensations are properties of oneself, it seems to follow that in bodily awareness one is aware of oneself. Bodily awareness, on this view, is a form of self-awareness. Thus, those who wish to deny that there is any sense in which we are experientially presented with ourselves can only do so by overlooking the case of our own bodies. We are embodied subjectivity and the lived body is a bodily self.

This line of thought might be challenged in a number of ways. First, it might be denied that pains and other bodily sensations really are located in the body. The view, introduced in §2.1, that bodily sensations involve perception of the body can be fleshed out in two ways. On the first, pains are properties of the body part in which they are felt. Thus, to have a headache is to perceive an ache in one's head. This is the view on which Brewer's account relies.

On the second, however, pains are the experiences of body parts as having certain properties (for example, tissue damage), rather than the located properties themselves. As we saw in Chapter 2, Brentano takes the latter view, arguing that it is a mistake to think of pains as literally located in body parts. Pains, being intentional, are not *located in* but rather *directed towards* the lived body. If this is correct, then we can challenge a key premise of Brewer's argument for the claim that the lived body – as location of sensation – is a bodily self.

Second, even if one is unimpressed by this objection and does agree with Brewer that pains and other bodily sensations really are located in the body, one may challenge the claim that sensations, so conceived, are properties of oneself. For, on this view, pains are entities that we are aware *of.* As such, it might be argued, there is some distance between myself and my bodily sensations. The perceptual awareness that I have of my pains may be a property of myself, but it is not obvious why we should think that the pain itself is.

Of course Brewer's argument, as represented above, is pitched at the level of metaphysics not phenomenology. What primarily concerns the phenomenologist is not whether sensations *are* located in the body, or whether they *are* properties of oneself, but whether they are *presented as such.* The conclusion with which the phenomenologist ought to be primarily concerned is not whether the body *is* the self but whether the lived body is *experienced* as oneself. Thus, a revised argument would run as follows: if bodily sensations are given as located properties of one's lived body and, further, bodily sensations are presented as properties of oneself, then bodily awareness is an awareness of one's body *as oneself.*

Restated in this way, the earlier objections might seem to have less force. First, whilst one might deny that pains *really are* located in one's body, we may still wish to say that they are *presented as* so located. My headache seems to be in my head, my toothache in my tooth, and so on. The body, to adapt Merleau-Ponty's phrase, is a 'sensation-space'. Second, even if one denies that sensations *really are* properties of oneself, one might nevertheless accept that they are *given* as such. In support of such a view we might point to the phenomenon of immunity to error, mentioned earlier. It seems to make no sense to wonder whether the headache that I am currently experiencing really is mine. One explanation for this would be that it is manifest in the experience itself that the headache is a property of me. The claim that the lived body is given as oneself might therefore be defended in this way.

3.3 Fusion

Regardless of exactly how the lived body is presented, it seems clear that one *can* be aware of one's own body as a thing. Through vision, as we have seen, one's body is given, in some respects, as are the bodies of others. Of course, as already noted, even through vision one's body is strikingly different from all other bodies. It is, for example, impossible to get the whole thing in view (at least without the aid of a series of mirrors). In fact, Merleau-Ponty goes further than this, claiming that aspects of the seen body are actually more akin to the lived body than to outer things, i.e. not given in a thing-like way. He writes:

> My visual body is certainly an object when we consider the parts further away from my head, but as we approach the eyes it separates itself from objects and sets up among them a quasi-space to which they have no access.
>
> (1945, p. 94)

The edge of the visual field is my body as seen. But it is unlike other things, even my own limbs, and it is so in a radical way. I cannot take a different perspective on it; I cannot remove it from view; it acts as a 'frame' allowing other things to be seen. In short – and taking on Merleau-Ponty's character-isation of what it is to be presented as a thing – just like the awareness of the lived body from the inside, in ordinary (non-mirror) visual experience my face seems not to be given as a thing.

  This is all apt to seem puzzling, however. On the one hand we have an inner awareness of one's lived body as the location of sensation, and so on. On the other hand we have an outer awareness of one's body as a thing among others. And, if Merleau-Ponty is to be believed, there are aspects of our bodily experience that defy this neat categorisation: an outer awareness of the body as lived. But surely we do not seem to ourselves to have two (or more) bodies. Our embodied experience is unified, meaning that we seem to ourselves to have just the one body, apprehended in distinct ways. How exactly are we to characterise this identification of the lived body with the body as thing? What, in other words, grounds our appreciation of the lived body as a thing among others?

  Broadly speaking, two options suggest themselves. The first would be that one identifies one's lived body with a certain outwardly presented thing in thought only. Noting that one's experience of the position, movement, and so on, of one's lived body is associated with the position and movement of that seen body that one cannot see the back of, etc., one naturally judges that the two are in fact identical. The second option is to regard such 'fusion' as taking place at the level of perceptual experience itself, that one's *Körper* is somehow also given as *Leib*, and vice versa. Stein defends this latter view:

> I not only see my hand and bodily perceive it as sensing, but I also 'see' its fields of sensation constituted for me in bodily perception [...]. The seen living body does not remind us it can be the scene of manifold sensations. Neither is it merely a physical thing taking up the same space as the

living body given as sensitive in bodily perception. It is given as a sensing, living body.

(1917, pp. 44–5)

According to Stein this is simply a special case of the same intermodal experience that we see in the awareness of outer things. As suggested in the previous chapter's discussion of Sartre's account of the unity of consciousness, the table that I see is *given as* the table that I feel. Further, the features that are presented in one sense modality are co-given in others. Thus, in Stein's words, '[w]e not only see the table and feel its hardness, but we also "see" its hardness. The robes in Van Dyck's paintings are not only as shiny as silk but also as smooth and as soft as silk' (1917, p. 44).

On this view, then, my visual experience presents my body as a thing but it also co-presents it as lived. Similarly, in bodily awareness my body is given as lived – as embodied subjectivity – but it is also co-presented as a thing – as the one thing I constantly see. This will be crucial in the next chapter as we consider Husserl's and Stein's accounts of the perceptual experience of other people in which, whilst proceeding *via* an awareness of their bodies, they are given precisely *as people*.

3.4 The vehicle of being-in-the-world

Much of the discussion in this chapter can be summed up with Merleau-Ponty's remark that the body is 'the vehicle of being-in-the-world and, for a living being, having a body means being united with a definite milieu, merging with certain projects, and being perpetually engaged therein' (1945, p. 84). According to Merleau-Ponty, the lived body is that which situates the self in the world, enabling us to actively engage with the things around us. Implicated in the experience of things it is not itself – not primarily at least – presented as a thing. It is, he tells us, prior to objective experience. This comes out in his critique of the 'objective thought' of 'the psychologist':

> Before being an objective fact, the union of the soul and the body thus had to be a possibility of consciousness itself [...]. To concern oneself with psychology is necessarily to encounter, beneath the objective thought that moves among ready-made things, a primary opening onto things without which there could be no objective knowledge. The psychologist cannot fail to rediscover himself as an experience, that is, as an immediate presence to the past, the world, the body, and others, at the very moment he wanted to see himself as just one object among others.

(1945, p. 99)

This 'primary opening onto things', of course, recalls Heidegger's original account of being-in-the-world which was, however, described without an explicit focus on the body. This is a lacuna that Merleau-Ponty sees himself as filling. The experience of one's own body and the experience of the world that it makes possible is in this sense 'pre-objective' (1945, p. 82). It is that which, according to Merleau-Ponty, enables the experience of objectivity – of experiencing things as things – and thus we must reject any approach to the body that attempts to treat it as an object just like any other. For Merleau-Ponty, embodiment is the fundamental grounding point for our experience more generally and the body is 'the pivot of the world' (1945, p. 84).

4 Conclusion

The fact that we are embodied is, of course, something familiar to us all. But the continuous awareness that each of us has of our own bodies 'from the inside' is something that we are apt to overlook. We are aware of our own bodies in a way in which we are aware of nothing else: as sensitive, as the centre of experiential space, and as immediately subject to the will. More than this, there is a case to be made that the body is a pre-condition of any form of awareness of things as things. The importance of embodiment for an account of our experience more generally cannot be overstated, a fact made vivid in the phenomenological accounts of Husserl, Stein, and Merleau-Ponty.

9   Experiencing others

The countenance is the outside of sadness. Together they form a natural unity.

Stein, *On the Problem of Empathy*

The previous chapters have, for the most part, proceeded as though experience were an entirely solitary affair. For the vast majority of us, of course, this could not be further from the truth. In fact we interact with other people on a daily basis and in a variety of ways: we see people act, sometimes joining in; we communicate with people, talking and listening in turn; we witness the ups and downs of people's lives, sympathising or being otherwise emotionally moved by their struggles and triumphs. Our lives are saturated with interactions with other human beings who, except in special circumstances, we treat as *people* rather than as mere things. Not only is this experience that we have of other people deeply significant for us, it also poses a great phenomenological challenge. For, as we have noted in our own case, people seem in some sense to be embodied subjectivity, unities of the experiential and the thing-like. But whilst it seems natural to say that we experience things through perceptual experience, and that we experience our own experiential lives through reflection, it is less obvious how we can describe the way in which the subjective life of another person is given to us. How, in our engagements with those around us, are their various experiences made available for us as objects? In this chapter we will focus on the account of how others are given that is presented by Husserl and Stein. As will be seen, far from being a self-contained phenomenological investigation, the consideration of the experience of others raises fundamental questions about any experience of Objects.

I The epistemological problem of other minds

When philosophers talk about the problem of other minds they usually mean the project of refuting a certain sort of sceptic (Avramides, 2001, Part I). We typically take ourselves to know that people other than ourselves exist. The sceptic about other minds challenges our right to claim such knowledge. There are a number of different ways in which such a sceptical argument can be mounted but, in very rough outline, one runs as follows: since the mental is private, we are privy to our own experiential life. That is, when we observe another human we perceive only outward signs of mentality, not the mentality itself. Thus, the belief in other minds is a mere prejudice, unfounded in experience.

As arguments go, this is not especially compelling. A very well-known response to it – the argument from analogy – maintains that since we are aware, from our own case, of a correlation between bodily behaviour and mental states and, further, we can observe similar behaviour in other humans, we are justified in believing that such a correlation also holds for those others. That is, we are justified in believing that they have minds. The classic formulation of this argument is presented by Mill:

> I conclude that other human beings have feelings like me, because, first, they have bodies like me, which I know, in my own case, to be the antecedent condition of feelings; and because secondly they exhibit the acts, and other outward signs, which in my own case I know by experience to be caused by feelings [...]. We know the existence of other beings by generalization from the knowledge of our own.
>
> (1872, pp. 243–4)

A familiar response to the argument from analogy is to object that such an inference is too weak to justify the belief in other minds since any correlation between behaviour and mentality is available to me only from my own case and we cannot take seriously an inductive inference based on such a small sample. Our belief, therefore, that other humans have a mental life cannot qualify as knowledge.

Whether Mill or the sceptic wins the day here is not our present concern. The aim of the argument from analogy is to support the commonsense claim that we know that people other than ourselves exist. This is evidently an epistemological rather than a purely phenomenological project. For our purposes, what matters are two claims that Mill and the sceptic seem to agree on. These are, first, that when we observe another human we are presented not with their mental life but with what we can call 'mere behaviour' and, second, that consequently our belief in the existence of the minds of others must be based on an inference from that behavioural evidence. These two claims fall, at least on one reading, within the province of phenomenology.

1.1 *Mere behaviour*

Suppose that you see someone step barefoot on an upturned pin. They will, in all likelihood, react in a predictable way. In seeing them react, we may ask, what is it that one sees? On the view outlined by Mill, what I see are 'modifications of [the] body', and 'outward demeanour' (1872, p. 243). The latter is what is often, in the contemporary philosophical literature, termed 'pain-behaviour'.

We get some indication of what such pain-behaviour is supposed to be when we consider that, according to Mill, we need to make an inference from it to the existence of pain. When I see someone exhibiting pain-behaviour, my subsequent judgement that they are in pain must be based on an inference, one that rests on an awareness of a correlation between my own pain-behaviour and sensations of pain. Pain-behaviour, then, is what we might call 'mere behaviour'. Whilst typically described in terms of the relevant sensation (pain), strictly speaking to ascribe mere behaviour is to ascribe something that falls short of mentality. To see someone's mere pain-behaviour is not to see any mental condition (pain) that they are in.

The claim that, in the observation of other humans, we are aware only of mere behaviour is, it would seem, a phenomenological claim. It is a characterisation of the *experience* of others. But the claim can seem to be motivated on grounds that are not strictly phenomenological. One is the plausible enough thought that experiences are not identical to any bodily movement. It is one thing to grimace, another to feel pain; one thing to smile, another to experience joy. Since, in the observation of others, we are unquestionably aware of bodily movements, it can seem natural to suppose that we are not aware of experiences. Another related thought is that it is possible for people to fake it. A skilled actor can behave in such a way as to be indiscriminable from someone in pain, and they can do so without themselves being in pain. Since, in the observation of an actor and a genuinely pained person, one is aware of the same thing, what one is aware of must be mere behaviour, not the pain itself. We shall return to these two lines of thought in §3.

1.2 Inference

The suggestion that our judgements concerning the mental lives of others are based on inferences can be interpreted epistemically, psychologically, or phenomenologically. Since these are not always carefully separated, it is worth explicitly distinguishing them. On an epistemological reading, the sorts of inferences to which Mill appeals in the above quotation need not characterise our actual movements of thought in the circumstances of observing and making judgements about others. Rather, they describe rational reconstructions, the suggestion being that our judgements are justified because they *could* be based on such inferences. Given its focus on justification, not phenomenology, for our present purposes we can set this reading to one side.

On a psychological reading, such inferences are ones that we actually make. That is, we form the judgement that another is in pain based on the perception of her pain-behaviour alongside the belief, gleaned from our own case, that pain-behaviour is correlated with pain. At first glance this suggestion may seem to fall within the province of phenomenology, since whether or not we engage in such inferences may seem open to phenomenological assessment. It is not entirely clear that this is so, however, since the notion of a non-conscious inference seems both to make sense and to fall outside the scope of phenom-enology. As Mill puts it, 'we may fancy that we see or feel what we in reality infer' (1891, p. 4).

What we are really concerned with is a phenomenological reading, according to which our judgements of the mental lives of others are *given as* based on inferences from mere behaviour. On this understanding, the view would be that the inferential nature of judgements about others' mental lives is accessible by way of reflection on our own experience. It is, as we shall see, the combination of this view, and the associated idea that we are aware only of the mere behaviour of others, to which Husserl and Stein object.

2 The co-presentation view

The alternative to the above picture is to maintain that, in observing another human, we are not aware of mere behaviour but of the other's experiential life itself. This is a view endorsed by Husserl (1931, Fifth Meditation), Stein (1917), and Scheler (1923, Part 3) (for contemporary versions, see Gallagher, 2008; Smith, 2010). As Stein puts it, for the case in which an emotion is seen in another's face, '[t]he countenance is the outside of sadness. Together they form a natural unity' (1917, pp. 76–7). If, when we observe another's sad countenance, we seem to be presented not with mere behaviour but with their sadness itself, then it would seem that there is no room for any sort of (phenomenologically salient) inference to the judgement that they are sad. Rather, if I judge that they are sad this will simply be a matter of taking the way they look at face value. And, as with visual experience, so with auditory experience since, according to Stein, '[c]heerfulness or sorrow, calmness or excitement, friendliness or rejection can lie in the tone of the voice' (1917, p. 78).

On this view, then, we are perceptually aware of others not simply as bodies exhibiting *mere* behaviour, but as persons displaying *psychologically rich* behaviour. As Scheler is at pains to point out, this is in fact the view of commonsense:

> we certainly believe ourselves to be directly acquainted with another person's joy in his laughter, with his sorrow and pain in his tears, with his shame in his blushing, with his entreaty in his outstretched hands, with his love in his look of affection, with his rage in the gnashing of his teeth, with his threats in the clenching of his fist, and with the tenor of his thought in the sound of his words. If anyone tells me that this is not 'perception', for it cannot be so, in view of the fact that a perception is simply a 'complex of physical sensations', and that there is certainly no sensation of another person's mind nor any stimulus from such a source, I would beg him to turn aside from such questionable theories and address himself to the phenomenological facts.
>
> (1923, p. 260)

This position is a version of what has become known in contemporary discussion as the 'high-level content' or 'liberal' view of perceptual experience, according to which perceptual experiences present not just 'low-level' properties such as shapes and colours, but also 'high-level' properties such as natural kinds or moral properties (Siegel, 2006). A complete description of our perceptual experience of others will need to employ notions such as 'sad', 'cheerful', 'excited', and so on. But a great deal more than this can be said. In fact, Stein presents an elaborate account of the experience of others that ties it directly to the Husserlian view of the experience of things and to her own account of the experience of embodiment. As we shall also see, however, there are a number of ways in which the account can be challenged.

2.1 Co-presentation and pairing

The key to the Husserlian account of the experience of others is the idea that we are co-presented with their mental life (for discussion, see Mooney, 2007). This is, in fact, a natural extension of the claim discussed in Chapter 8, that the awareness that one has of one's own lived body is 'fused' with the awareness that one has of it as a thing (we will return to fusion in §3.2). As we saw, the view is that outer experience, such as vision, presents my body as a thing but also co-presents it as lived. The awareness of the bodies of others is no different, according to Stein:

> we have a primordial givenness in 'bodily perception' of our own fields of sensation. Moreover, they are 'co-given' in the outer perception of our physical body in that very peculiar way where what is not perceived can be there itself together with what is perceived. The other's fields of sensation are there for me in the same way. Thus the foreign living body is 'seen' as a living body.
>
> (1917, p. 57)

In the perceptual experience of a thing one is aware of its facing side with intuitive fullness and its hidden sides in an empty way. They are co-given or, to use an expression that both Husserl and Stein employ in this context, they are not 'original' or 'primordial'. This combination of givenness and co-givenness is what it is to be aware of an entity as a thing. In the perceptual experience of another human one is, in this way, aware of their body as a thing. However, additionally, their mental features are co-given and this is what it is for them to be given *as a person*. Thus, in the observation of another human, one is not aware of mere behaviour and so required to *infer* that they possess a subjective life, rather one is aware of their experiential life by way of one's perceptual experience itself.

Whilst this may strike us as an apt way of describing the experience of others, it is very natural to ask why it is that such co-givenness occurs. Why, that is, would it be that my experience of other humans co-presents them as people whereas my experience of rocks and trees does not? In *Cartesian Meditations* Husserl does in fact offer an answer to this question, one that relies on the fact that other people's bodies, by and large, resemble our own. Indeed, his answer is in some ways reminiscent of the argument from analogy (see Ricoeur, 1967, for discussion).

Husserl claims that some feature of experience must 'motivate' the co-presentation of another's body as lived and the account that he gives draws on his notions of 'analogising transfer' and 'pairing'. On Husserl's picture, when I see something *as a thing of a certain kind*, this is either a 'primal instituting' or it involves an 'analogising transfer' from a primal instituting (1931, p. 111). A primal instituting is the first instance of seeing something as being of that kind. Subsequent experiences of something that appears similar refer back to the original and there is an 'analogising transfer' of the sense of the primal instituting to the new case:

> *each everyday experience* involves an *analogising transfer* of an originally instituted objective sense to a new case, with its anticipative apprehension of the object as having a similar sense [...]. The child who already sees physical things understands, let us say, for the first time the final sense of scissors; and from now on he sees scissors at the first glance *as* scissors.
>
> (1931, p. 111)

Analogising transfers are particularly evident in cases of pairing. When I simultaneously experience two objects that appear similar, I see them *as a pair*. In such cases, the sense attached to the one is analogously transferred to the other. That is, each appears *like the other*: 'we find [...] a living mutual awakening and an overlaying of each with the objective sense of the other' (1931, p. 113).

In the case of the experience of another person, Husserl claims that we experience our own bodies as paired with the bodies of others. This pairing is grounded in the fact that the bodies of others appear similar to our own bodies. Further, as we saw in the previous chapter, our own body is experienced by us in a particular way. We are continually aware of our own body as a *lived body*. It is given as the location of sensations, as at the centre of perceptual space (as 'here'), as the immediate object of my will, and as that in which my emotions are expressed. Consequently, according to Husserl we experience the other's body *as a lived body*. We do not, of course, experience the other's body as lived in the same way that we experience our own as lived. Rather, since we experience it as paired with our own lived body, the sense *lived body* is 'analogously transferred' to the body of the other. Thus, the other's body is co-presented as lived:

> pairing first comes about when the Other enters my field of perception. I, as the primordial psychophysical Ego, am always prominent in my primordial

field of perception [...]. In particular, my lived body is always there and sensuously prominent [...] equipped with the specific sense of a lived body. Now in case there presents itself, as outstanding in my primordial sphere, a body 'similar' to mine – that is to say, a body with determinations such that it must enter into a phenomenal *pairing* with mine – it *seems* clear without more ado that, with the transfer of sense, this body must forthwith appropriate from mine the sense: *lived body.*

> (1931, p. 113, translation amended)

Notice that, in the above passage, Husserl says that it *seems* apparent that the sense *lived body* is transferred. This caution alludes to the question of why it would be that in experiencing a perceived body *as a lived body* it would be given as the lived body of another, 'rather than a second body of my own' (1931, p. 113). In being aware of a foreign body as lived, why am I not aware of it as an extension of my own?

Husserl's suggested response to this is that since the livedness of our own bodies is always *given* rather than merely *co-given*, a body that is merely co-given as lived must be given as the body of *another*, rather than as a second body of mine. For example, since my pain is always present, rather than co-presented, to me, the fact that I am aware of a body that is co-presented as in pain is just what it is for it to be experienced as the body, and pain, of another.

There is actually a deep puzzle here. What Husserl is grappling with at this point is the so-called 'conceptual problem of other minds'. This is the question of how it is that, if the concepts *ego, experience, sensation*, etc. are grasped *via* a reflection on my own case – myself, my experience, my sensation, etc. – these concepts nevertheless possess such generality as to allow them to be applied to others. That is, why, if I know what pain is from my own case, is my concept of *pain* not the concept of *my pain* and so any awareness of another as being in pain is, in truth, an awareness of them as having *my pain*, i.e. not really an awareness of them as *another* person at all? This conceptual problem of other minds (see Avramides 2001, Part III) is particularly associated with the work of Wittgenstein, who famously expressed it thus:

> If one has to imagine someone else's pain on the model of one's own, this is none too easy a thing to do: for I have to imagine pain which I *do not feel* on the model of the pain which I *do feel.*
>
> (1953, §302, p. 101)

Whether Husserl possesses an adequate answer to this puzzle is not a question that we can pursue here, although a version of it will resurface in the discussion of the problem of perceived similarity in §3.2. It must be borne in mind, however, that the conceptual problem is never far from the surface of any account of experiencing others.

So, the conceptual problem of other minds aside, Husserl and Stein present a view according to which others' bodies are given in experience as lived: as the location of bodily sensation, as at the centre of a distinct perceptual space, as under the direct control of someone other than ourselves, and as expressive of their psychological life. We shall briefly look at Stein's discussion of each of these below. In spelling them out, as we shall see, Stein is presenting an account of our perceptual awareness of other people as founded on a more fundamental level of experience – that of the experience of foreign bodies and of the experience of our own bodies with which we are familiar from Chapter 8.

2.2 Sensation, perception, action, and expression

On Stein's subtle account, the experience of others is structured, in the sense that some aspects of the full awareness of another person are based on others. On, or at least near to, the 'ground floor' are both the awareness of the other's 'fields of sensation', and the sensations themselves, and also the experience of their body as a center of perceptual orientation. Regarding the former, she writes:

> The hand resting on the table does not lie there like the book beside it. It 'presses' against the table more or less strongly; it lies there limpid or stretched; and I 'see' these sensations of pressure and tension in a co-primordial way.
>
> (1917, p. 58, translation amended)

The hand that I see on the table resembles my hand so I see it not simply as a lump of flesh exhibiting mere behaviour, but as a lived hand pressing on and thereby sensing the table.

As with fields of sensation, so with spatial orientation. Since the foreign body 'is spatial like other things, and is given at a certain location' (1917, p. 61), and since it resembles my own body, which is experienced as located at the centre of orientation, so the other's body is co-presented as the origin of 'a new zero point of orientation' (1917, p. 61). In this way, then, I am aware of the other's body as a lived, sensory body at the centre of a perspective on the world.

According to Stein, the awareness of a foreign body as the direct object of another's will – as something with which a person acts, rather than as something that merely moves around – depends on its already, in the above way, being co-presented as lived. The reason for this is that bodies are capable of mechanical or 'associated' movement, as when my arm is lifted by someone else, as well as spontaneous willed movement, as when I raise my arm. So,

> If I see someone ride past in a car, in principle his movement appears no differently to me than the 'static' parts of the car. It is mechanical associated movement [...]. The case is entirely different if, for example, he raises himself

up in the car. I 'see' a movement of the type of my spontaneous movement. I interpret it as his spontaneous movement.

(1917, p. 67)

This difference between 'associated' and 'spontaneous' movement is crucial, but not sufficient, to account for the difference between seeing something as moving and seeing it as acting. For all sorts of non-animal bodies move seemingly spontaneously, without thereby being given as acting in accordance with a foreign will. Rather, for the spontaneous movement of a body to be given as the willed action of another person, 'we must already have interpreted it as a living body' (1917, p. 67). To experience something as willed movement is to experience it as the spontaneous movement of a lived body.

Stein singles out the phenomenon of expression for special treatment. As I suggested in Chapter 8, expressive bodily activity sits somewhere between mechanical and willed movement. The movements of my face when I am afraid are not experienced as things that are simply happening to me, nor are they experienced as things that I am *doing* in the full sense. Rather, they are experienced as the natural bodily vehicles of an outpoured emotion. Similarly, on Stein's view, expressive bodily activity is not given to us in just the same way as are willed movements:

> feelings can also be co-comprehended together with movements. For example, I can see a person's sadness by his gait and posture. However [...] the movement is not sad in the same way that the countenance is sad. The sadness is not expressed in the movement. On the contrary, emotional expressions are on exactly the same plane as visible movements of expression. Fear is at one with the cry of fear just as sadness is with the countenance.
>
> (1917, p. 78)

Whilst the detail of this is perhaps somewhat obscure, it does seem clear that Stein wishes to distinguish between different ways in which we can become aware of the experiential life of others. The first, we have already encountered. The second, that which is made available through expression, is described in such a way as to minimise the distinction between that which is presented and that which is co-presented: 'fear is at one with the cry of fear'. Again,

> When I 'see' shame 'in' blushing, irritation in the furrowed brow, anger in the clenched fist, this is still a different phenomenon than when I look at the foreign living body's level of sensation [...]. In the latter case I comprehend the one with the other. In the former case I see the one through the other.
>
> (1917, p. 75)

This ought to put us in mind of the phenomenon of 'seeing-in' discussed in Chapter 6. When I look at a picture of a tree, I see a tree *in* the picture.

This, whilst not the same as seeing a tree in the flesh, is nevertheless a way of a tree's being present to me. Stein seems to be suggesting something similar with respect to the phenomenon of expression. Whilst, in the case of sensation or spatial orientation, I see a body which is co-presented as a sensory, zero-point of orientation, in the case of expression I see the expressed emotion *in* the face. The shame and the blushing are given as aspects of one 'experiential whole', made 'intelligible' by the fact that the former 'motivates' the appearance of the latter (1917, p. 84).

This is suggestive, but certainly in need of further elaboration. However, whilst there is clearly a great deal more to be said regarding the phenomenon of expression (for an insightful, non-phenomenological discussion, see Green, 2007), it is undoubtedly very natural to suppose that there is something special about the way in which facial and other expressions bring the other's sub-jective life into view. On Stein's view it qualifies as something like the pinnacle of our awareness of others, representing the most direct form of awareness that we have of their experiential lives, and resting as it does on the more fundamental awareness of them as bearers of sensation at the centre of their own perceptual space.

3 Is the co-presentation view credible?

There are, of course, a number of ways in which the above account can be challenged. I begin with two that can be reasonably swiftly dealt with, going on to consider two worries that are significantly less easy to dismiss, the second of which paves the way for the introduction of a piece of the puzzle that has been missing so far: empathy.

3.1 Mere behaviour again

In §1 I mentioned two lines of thought that might seem to motivate the claim that in observing another we are really only ever aware of mere behaviour. The first was that since there is no question that we are perceptually aware of the other's bodily behaviour and also that psychological states are not *the same thing as* bodily behaviour, we should agree that we are not perceptually aware of psychological states but rather of mere behaviour. That this con-clusion does not follow from the two premises is obvious once we see that the co-presentation view, whilst conceding that we do experience bodily behaviour, maintains that we *also* experience psychological states. The former is pre-sented to us and the latter is co-presented. Thus, on this view we are aware of behaviour *as* co-presenting the subjective, experiential life of the other person. This is clearly not *mere* behaviour.

The second line of thought started from the recognition that a skilled actor might be able to perfectly mimic expressive behaviour. Further, since in observing an actor and someone in pain, one is aware of the same thing, that thing must be mere behaviour, not the pain itself. This argument is obviously related to the arguments from illusion and hallucination discussed in Chapter 3, and the moves made there are applicable here also. Interpreted as a variation on the argument from hallucination, the proponent of an intentional account of experience will maintain that the line of thought goes astray in claiming that in the two experiences – of the actor, on the one hand, and of the person in pain, on the other – one is aware of just the same thing. On the contrary, in the experience of someone in pain, one is aware of their pain (by way of its being co-present), whilst in the experience of an actor, one is not aware of pain but one *seems to be*. That is, when one observes an actor playing the role of someone in pain, one has a non-veridical experience. The actor seems to be in pain, but is not. As Stein says: '[a]s in every experience, deceptions are here also possible' (1917, p. 86).

3.2 Perceived similarity

The co-presentation account of the experience of others relies on the claim that the other's body appears similar to my own lived body or, to put it another way, the claim that they are experienced as of the same type (human body). This is what grounds the analogising transfer of the sense *lived body*. As Stein says: '[t]he interpretation of foreign living bodies as of my type helps make sense out of the discussion of "analogizing" in comprehending another' (1917, p. 59). But, it can be objected, my body doesn't *look* all that similar to the other's body (cf. Schutz, 1970, pp. 63–4). As we saw in Chapter 8, my own body is presented from a fixed angle, has a large hole at the level of the head, and, if we are to agree with Merleau-Ponty, incorporates elements (e.g. my nose) that are not given as things at all.

Husserl is aware of this problem and proposes, as a solution, that one does not experience pairing between the foreign body and one's own body as it *actually* appears, but rather as it *would* appear were it *over there* where the foreign body is located. The other's body, as he says,

> does not become paired in a direct association with the manner of appearance actually belonging at the time to my lived body (in the mode Here); rather it awakens reproductively *another*, an immediately similar appearance included in the system constitutive of my lived body as a body in space. It brings to mind the way my body would look 'if I were there'.
>
> (1931, pp. 117–18, translation amended)

This may well appear an inadequate solution to the problem, however, since although there is a sense in which a foreign body does look like my own body would if I were over there, it doesn't look similar to how my body would look *to me* if I were over there. If I really were over there my body would look just as it does now! This is for the reason, discussed in some detail in Chapter 8, that the lived body is the zero-point of perceptual space. If my body were over there, then over there is where I would experience it from.

One suggestion to mitigate this problem would be to suppose that the foreign body looks similar to the way my own body would to *a person located here* if I were there. But that, it seems, would be circular. For such a view seemingly presupposes the capacity to represent the viewpoint of a person other than myself and that, of course, is exactly what Husserl is attempting to provide an account of.

In any case, it may be further pointed out, I am not aware of the livedness of my body *via* visual perception. Rather, I am aware of its livedness – its being the location of sensation, and so on – *via* bodily awareness; the experience of my own body from the inside. It would seem, then, that in order for there to be an experience of pairing between my body and a perceived foreign body – a pairing that would enable a transfer of the sense *lived body* – it must be the case that that body is experienced as similar to my own body, as given from the inside through bodily awareness. I am not, however, aware of the other's body *via* bodily awareness, but through the outer senses of vision, and so on. But, in what sense is my body as it appears *via* bodily awareness similar to the other's body as it is *seen*? It might well seem that it is not at all similar – the way that my body *feels* is, for me, very different from the way the other's body *looks*. If this is right then Husserl and Stein's claim that we experience the other's body as one of a pair with our own – and in such a way as to ground an analogising transfer of the sense *lived body* from my own to the foreign body – might be rejected.

As a solution to this worry we could, with Stein, appeal to the discussion of fusion from Chapter 8. There we saw that Stein claims that there is a fusion between my experience of lived embodiment from the inside and my experience of my body as an object from the outside. According to her, whilst it is only the experience of one's body from the inside that presents it as lived, it is nevertheless co-presented as lived even through outer awareness. In relation to the experience of others, we are told that the account

> is warranted by the interpretation of our own living body as a physical body and our own physical body as a living body because of the fusion of outer and bodily perception. It is also warranted by the possibility of spatially altering this physical body, and finally by the possibility of modifying its real properties in fantasy while retaining its type.
>
> (1917, p. 58)

Because of the fusion of inner and outer, there is a sense in which I am visually aware of my body as *lived*. And because I also have an awareness of my lived body as a thing, I can imagine it *over there*. So the experience of a foreign body can be paired with my body as it would look if it were over there, an experience in which it would itself be co-presented as lived.

Once more, however, this proposed solution can be challenged. To see this, recall that, in the case of the experience of a foreign body, Husserl maintained that some feature of experience must 'motivate' the co-presentation of another's body as lived. This, indeed, was the starting point of the discussion of pairing. But just as we asked what motivates the co-presentation of the other's body as lived, so we can ask what motivates fusion; what motivates the co-presentation of the seen body as lived and of the lived body as a thing such that one has an awareness of oneself as a unified psychophysical thing among things.

The difficulty with employing fusion to solve the problem of perceived similarity can be seen when we notice that both Husserl and Stein suppose that what motivates fusion is the awareness of the *other* as having a view on oneself as a body with co-present subjective life:

> From the viewpoint of the zero point of orientation gained in empathy, I must no longer consider my own zero point as the zero point, but as a spatial point among many. By this means, and only by this means, I learn to see my living body as a physical body like others.
>
> (1917, p. 63)

But if this is correct, then the fusion of the inner and outer experience of my body into an awareness of it as a unified psychophysical thing cannot be that which enables me to experience the pairing of my body with that of the other. The account, on this reading, would be circular.

In fact, the problem seems to go deeper than this since both Husserl and Stein accept the further view (mentioned in Chapter 7) that the experience of *anything* as objective presupposes an awareness of the perspective of others. It is in having an awareness of a thing as simultaneously experiencable from distinct perspectives that I have a sense of it as transcending my own experience of it. That is, the awareness of others is a condition of the possibility of the awareness of entities as Objects. Stein writes:

> The same world is not merely presented now in one way and then in another, but in both ways at the same time. And not only is it differently presented depending on the momentary standpoint, but also depending on the nature of the observer. This makes the appearance of the world dependent on individual consciousness, but the appearing world – which is the same, however and to whomever it appears – is made independent of consciousness [...] intersubjective experience becomes the condition of possible knowledge of the existing outer world.
>
> (1917, p. 64)

On this account, then, the awareness of others is more basic than the awareness of objective, worldly entities as such. A different route to a similar conclusion can be drawn from Heidegger's account of the role of social norms in intentionality. Heidegger no doubt regards Husserl's and Stein's accounts as 'theoretical concoctions' (Heidegger, 1927a, p. 116), untrue to the phenomena. Recall from Chapters 1 and 2 that, according to Heiddegger, intentionality is possible only because we possess a pre-ontological understanding of the being of entities. Furthermore, this understanding manifests itself in a practical capacity to engage with entities of the sort in question within a contextual whole that is *the world*. But more than this, claims Heidegger, we must recognise that 'the world is always already the one that I share with others. The world of Dasein is a *with-world*' (1927a, pp. 115–16). This, thinks Heidegger, has far-reaching consequences, 'the understanding of others already lies within the understanding of being of Dasein because its being is being-with. This understanding [...] makes knowledge and cognition possible' (1927a, p. 120). How is this so? How does the awareness of others make 'knowledge and cognition' possible? A short answer is given by Dreyfus:

> For eating equipment to work, how *one* eats, when *one* eats, where *one* eats, what *one* eats, and what *one* eats with must be already determined. Thus the very functioning of equipment is dependent upon social norms. Indeed, norms define the in-order-tos that define the being of equipment, and also the for-the-sake-of-whichs that give equipment its significance.
>
> (1991, p. 154)

In order for something to show up as equipment, one must understand it and that means that there must be some norm prescribing its use. These norms – of which we are implicitly aware in being aware of something as equipment – are social, they are an aspect of the with-world. Thus, Heidegger sees the awareness of others as fundamental (for discussion, see Hall, 1980; Olafson, 1994; Carman, 1994).

If the awareness of others is, in either of these two ways, prior to the awareness of things as objective, then this suggests a deep flaw in the proposed solution to the problem of perceived similarity. According to the solution, my own body is paired with that of the other because, due to fusion, I have an awareness of my body as a thing such that I can imagine how it would look, were it over there. This seems to depend on an awareness of it as a thing which can be ways other than those that it is in fact experienced as being, i.e. an awareness of it as an Object. But, according to the above suggestion that intersubjective experience is in fact a condition of the experience of anything as an Objective thing (a thing transcending the experience we have of it), such an awareness of one's body as an Object depends on, and so cannot ground, the experience of others.

3.3 Anticipation

A different challenge to the co-presentation account focuses not on the account of pairing but on the co-presentation claim itself. It is clear, this challenge runs, that there is a significant disanalogy between the way in which I am aware of the non-facing sides of things, on the one hand, and the subjective states of other people, on the other.

As discussed in Chapter 3, Husserl uses the term 'anticipation' to describe the way in which the co-presented aspects of a thing are given in perceptual experience. What is anticipated are further experiences, contingent on my or the thing's movement. So, as I see the tree outside my window, its rear side is co-present by way of the anticipation of its being fully presented were I to walk around it, for example. On this understanding of co-presentation, for something to be co-presented it must be possible – or at least must seem possible – to bring it into full view.

As a thing, the same can be said for a foreign body in my field of view. It will be seen from a particular perspective with its occluded sides merely co-presented, so perceptually anticipated. But it seems that the same cannot be said for the supposedly co-presented livedness – location of sensation, zero-point of orientation, etc. – of a foreign body. For there is no perspective that I can occupy which would bring these features into full view. There is no possibility of their ever being anything other than co-presented. Husserl is familiar with this concern:

> experiencing someone else cannot be a matter of just this kind of appresentation [co-presentation] [...]. Appresentation of this sort involves the possibility of verification by a corresponding fulfilling presentation (the back becomes front); whereas, in the case of that appresentation which would lead over into the other original sphere [the other person's experiential perspective], such verification must be excluded a priori.
>
> (1931, p. 109)

This leaves the view with a particularly pressing difficulty. Obviously enough, the co-presentation account crucially rests on the notion of co-presentation. But, it seems, the concept of co-presentation is typically described in such a way as to make it implausible that it could perform the role allocated to it in the account.

Husserl's own response to this worry is that what is perceptually anticipated in the case of co-presented aspects of the other's psychological life are further presentations of bodily behaviour that co-present appropriately related experiential features. For example, if I see a body with eyes open, facing a tree, my perceptual experience will co-present it as seeing the tree from a certain angle. What this involves is the perceptual anticipation of an experience which will co-present the body as seeing the tree from a different angle were it to move appropriately. In Husserl's words:

> Regarding experience of someone else, it is clear that its fulfillingly verifying continuation can ensue *only by means of new appresentations that proceed in a synthetically harmonious fashion.*
>
> (1931, p. 114)

The co-presentation of another's mental life, on this suggestion, involves their 'changing but incessantly *harmonious behaviour*' (1931, p. 114).

Whilst this does give some sense to the key notion at work in the co-presentation account, it is apt to seem unsatisfactory. For it may be complained that the account, construed in this way, just gives us more behaviour where what we wanted was an experience of the other's mental life. The experience of another's mental life, on this picture, really only amounts to the anticipation of more behaviour. This, it might seem, is insufficient to ground a full appreciation of the other's perspective, since it grants us no sense of *what the other's experiential life is like*. That is, we still require some account of how it is that we have an awareness not only of foreign bodies behaving in various anticipated ways but also of *how things are* for a *subject of experience*. Fortunately for those who favour the co-presentation account but also take this objection seriously, there is a response: empathy.

4 Empathy

The concepts of empathy and 'fellow-feeling' are important within the Phenomenological tradition (see Moran, 2004; for a classic discussion see Scheler, 1913/16, Part 1, Ch. 2). In the present context, we are interested in the role that empathy plays in the experience of others and, in particular, whether an appeal to empathy can make intelligible to us how the co-presentation of another's mental life can allow us to appreciate *how things are* for another person. That Stein takes empathy to play such a role is clear (for discussion, see Meneses and Larkin, 2012). Speaking of the tendencies to anticipate further experience that are involved in the co-presentation of the rear aspects of things, she writes:

> The co-seeing of foreign fields of sensation also implies tendencies, but their primordial fulfillment is in principle excluded here. I can neither bring them to primordial givenness to myself in progressive outer perception nor in the transition to bodily perception. Empathic representation is the only fulfillment possible here.
>
> (1917, p. 57)

The question is what empathy might be so as to perform this role. Unfortunately, empathy is a widely contested concept of which there are a number of distinct construals. Most accounts, however, involve some combination of imagination – or 'perspective-taking' – and vicarious experience. As such, a simple understanding of empathy is of a process of imagining being in another's situation and, as a result, in some sense feeling as they do (cf. Coplan and Goldie 2011).

Talk of imagination brings to mind an aspect of Husserl's account of pairing that we have already discussed – the suggestion that the foreign body is paired with how my own body would look were it over there. This is for the reason that it is natural to suppose that I have some sense of how my body would look if over there because I am able to imagine it as such. Whilst Stein does claim that empathy is in some ways like a certain sort of imagination, in fact she does not think of empathy in quite this way. It is similar in that it involves a 'non-primordial' awareness of an experience. When, for example, I imagine being another person – when I 'put myself into their shoes' – I might have an imaginative awareness of having various experiences. I do not, however, *actually* have those experiences. That is, they are not given to me *in the flesh*, rather I intend them in an empty way. Empathy, similarly, involves an empty intending of another's mental states. However, it is distinct from the imagination since the experiences that one is aware of in imagination 'are not given as a representation of actual experiences' (Stein, 1917, p. 9). In empathy, on the other hand, they are. One empathises with the other's actual experience.

Stein would reject any account of empathy that requires the empathiser to actually undergo an experience similar to that undergone by the person with whom they empathise. The simple account of empathy mentioned above states that empathy involves 'in some sense' feeling as the other does. On Stein's account, one does not literally feel as the other does, but rather has an empty representation of the way they feel. Distinguishing empathy from first-hand experience, from imagination, and from memory, she writes that

> while I am living in the other's joy, I do not feel primordial joy. It does not issue live from my 'I'. Neither does it have the character of once having lived like remembered joy. But still much less is it merely fantasized without actual life.
>
> (1917, p. 11)

This tells what empathy is not, but what positive description does Stein give of empathy, and how is it to play the role of bringing home to us how another feels? Stein's account begins with the co-presentation of the other's mental state. In such an experience, I am aware of the other's mental state 'as an object'. But, she goes on:

> when I inquire into its implied tendencies (try to bring another's mood to clear givenness to myself), the content, having pulled me into it, is no longer really an object. I am now no longer turned to the content but to the object of it, am at the subject of the content in the original subject's place.
>
> (1917, p. 10)

Consider again the example of seeing another person seeing a tree. I experience their body as facing a tree and this co-presents their seeing a tree. This, if we accept Husserl's account, involves the anticipation of further experiences presenting the body and co-presented related seeings. When, however, I undergo this experience, I am drawn into it and find myself as if occupying the other's position and myself seeing a tree. Stein also describes this 'being drawn in' as the more active sounding experience in which I 'project myself into' the other's experience. Either way, of course, I don't really see the tree from their perspective, but it is *as if* I do. As a result, I have an empty intending of their experience, but one that gives me a sense of how things are for them; of how the world appears from their perspective.

If these descriptions are accurate, then, we have in empathy an answer to the outstanding worry about the nature of the co-presentation of another's mental life. As Stein puts it: '[t]he experience which an "I" has of another "I" as such looks like this. This is how human beings comprehend the psychic life of their fellows' (1917, p. 11).

5 Conclusion

The experience of others is a perennial topic in both epistemology and phenomenology. How exactly we describe such an awareness, and what Heidegger calls the with-world, is highly significant, especially if we find plausible the idea that intersubjectivity is somehow implicated in any experience of Objects. The account defended by Husserl and Stein can seem phenomenologically compelling in some respects (our awareness of others is perceptual not inferential) but less so in others (the exact account of pairing). Since sociality is an absolutely ubiquitous aspect of our experiential lives, it is important to get its phenomenology right and so determine exactly what is involved in the experience of others.

10 Experiencing emotion

emotion arises when the world of the utilizable vanishes abruptly and the world of magic appears in its place

Sartre, *Sketch for a Theory of the Emotions*

An account of human experience would be deficient if it failed to take note of the fact that we are emotional beings. Not only do we see, hear, touch, imagine, and act, we also feel good or bad and do so in a number of ways. Indeed, the experience of emotion is one of the most salient and important features of our day-to-day lives. I awake from a dream with a paralysing feeling of terror. As mid-life approaches I view my past and future with a pervading sense of anxiety. As yet another petty demand on my time arrives via email, I experience a rising anger. On seeing my daughter's first smile I am overcome with unconditional love. Now grown, I am buoyed with pride as I watch her roller-skate for the first time. In this chapter we will look at the nature of such emotional experience considering, amongst other things, its relation to bodily experience and introducing for the first time in this book our relation to the world of value.

1 Emotional experience

Our topic is emotional experience and, since this is not obviously the same thing as emotion itself, it is worth saying something now to delimit that with which we will be concerned. If a volcano erupts in my vicinity, posing an immediate existential threat to myself and those around me, I am likely to feel fear. This is an experience and thus falls within the domain of a phenomenological investigation. This, however, is not true of every fear of volcanoes. If I have a longstanding fear – or even phobia – of volcanoes, this need not manifest itself in my experiential life at every moment at which it exists. It is, for example, true of me that I am afraid of volcanoes even while asleep or while entirely absorbed in some non-volcano related activity. This sort of fear, then, is not itself an experiential phenomenon and so is not our immediate concern. This division is sometimes marked by distinguishing between episodic and dispositional emotions and the claim that only the former constitute elements of our experiential lives (Deonna and Teroni, 2012, pp. 7–11). In these terms, it is episodic emotions with which we are concerned, although I shall typically use the phrase 'emotional experience' or simply just 'emotion'. In the remainder of this section I will briefly introduce some reasonably widely accepted features of emotional experience. We can then move on to consider how Sartre's contribution to the topic fits in to this broad picture. For it is Sartre's early work on emotion that will provide the central point of discussion.

1.1 Intentionality

In Chapter 2 we considered Brentano's view that all and only mental phenomena exhibit intentionality. As experiential episodes, emotions are evidently mental phenomena and so on this view must be intentional. Indeed, that emotions are intentional is both a part of our commonsense conception of emotional experience and a commonplace observation in philosophical discussions (see Deonna and Teroni, 2012). If I am afraid then there is something of which I am afraid: the fierce dog, the ghost, falling over the edge, failing my exams, and so on. If I feel joy, then there is something about which I feel joy: the landscape, passing my driving test, the glory of God, and so on. As Husserl puts it in *Logical Investigations*, '[i]t seems obvious, in general, that every joy or sorrow, that is joy or sorrow *about* something we think of, is a directed act' (1900–1, Vol. 2, p. 107). Husserl follows Brentano in claiming that emotions are not themselves presentations but rather are *founded* upon presentations (recall the brief discussion in Chapter 2). That is, one can only undergo an emotional experience directed to something if that thing is already presented to one in, say, perception or memory. Nonetheless, the emotion itself is intentionally directed:

> Whether we turn with pleasure to something, or whether its unpleasantness repels us, an object is presented. But we do not merely have a presentation with an added feeling *associatively* tacked on to it, and not intrinsically related to it, but pleasure or distaste *direct* themselves to the presented object, and could not exist without such a direction.
>
> (Husserl, 1900–1, Vol. 2, pp. 107–8)

If emotional experiences have intentional objects then we would expect them to possess intentional content. That is, emotions are not simply directed at objects but at experiences in which objects are given as being some way or other. Exactly how emotions present their objects is a matter of controversy but a common thought is that it involves an awareness of what we might call 'affective qualities'. Thus, in fear I experience the feared object *as frightening*, in joy I experience an object *as wonderful*, and so on. We shall return to this issue below when considering the relation between emotional experience and evaluation.

Emotions were mentioned in Chapter 2 for the reason that some emotional phenomena – moods such as depression or anxiety – are considered by some philosophers to be counterexamples to Brentano's claim that all mental phenomena are intentional. There I pointed out that, on a broadly Heideggerian conception of moods, they do not lack intentionality but are rather directed at something very general, one's being-in-the-world as such. We shall return to this long-postponed issue in §3.

1.2 Feeling

One of the most striking features of emotional experience – already indicated in the above quotation from Husserl – is that when one undergoes some emotion, one *feels* a certain way. Emotions have, as we might say, an *affective* character, I am affected by them. And this is reflected in the way in which we speak of emotion: we say not just that I fear the ghost but that I *feel* afraid of it, not only that I hate Nigel but that I *feel* animosity towards him.

There are a number of different ways in which one might think about the affective character of emotional experience. Perhaps the most elementary thought is that emotions seem to have a *valence*: some emotional experiences feel bad whilst others feel good. For example fear, hatred, and anxiety have a negative valence (they feel bad), whereas joy, love, and pride have a positive valence (they feel good).

It is generally recognised, however, that there is more to the way that emotional experiences feel than simply their valence. For one thing, emotions with the same valence, for example fear and hatred, can feel different. For another thing, a good number of emotional experiences seem to possess a bodily phenomenology. Thus, in being afraid of a ghost I feel the hairs on the back of my neck stand on end; in experiencing a sudden surge of anger I feel my heart beating violently and the blood racing in my veins; in experiencing joy I feel a warm glow and sense of ease spread throughout my body. Emotional experience is, at least in part, bodily experience.

1.3 Evaluation

We experience emotions in relation only to things about which we care; things that matter to us. This much seems obvious with positive emotions such as love, since loving something seems to be a paradigm case of caring for it. But it is equally the case with negative emotions. I fear the ghost because I care about my well-being, hate Nigel because he offends my sense of decency and justice, things about which I care, and so on.

This relation between emotion and care brings us back to the suggestion that there is an important connection between emotion and evaluation. Emotions seem to either be, or be dependent on, an evaluation of their objects. Thus, if I hate something I find it offensive, if I fear something I consider it threatening, if I am disgusted by something I take it to be repulsive, and so on. Such evaluations, it would seem, are directed toward both the valued thing and the value itself. As Husserl puts it: 'If we are directed to a thing in an act of valuing, then our direction to the thing itself is a heeding of it, a seizing upon it; but we are "directed" [...] also to the *value*' (Husserl, 1913, p. 77; also see 1913, §95).

A claim sometimes made about such evaluations is that they serve to distinguish emotional experiences from each other. So, the difference between love and pride, between hate and fear, and so on, is ultimately a matter of how each evaluates its object. This is sometimes put by saying that such affective features are the 'formal objects' of the emotions and that the essence of the individual emotions is the evaluation of intentional objects as possessing them (Kenny, 1963).

1.4 Expression and action

The expression of emotion was mentioned in Chapters 8 and 9 as constituting an aspect of the lived body. Recall that Stein claimed that I experience my body as that in which emotions terminate by way of an 'outpouring' of affective experience. My joy, for example, tends towards expression in a smile, the constitutive movements of which I am aware of from the inside. Such expressive behaviour, then, can be added to the bodily feelings mentioned above, adding to the sense that emotions possess an experiential character that is, at least in part, bodily.

Alongside bodily changes, such as the racing heart, and expressive behaviour, such as smiling, it is also natural to think of emotional experience as associated with tendencies to act in various ways. So, when afraid I will be disposed to flee, when feeling love I will be disposed to embrace, and so on. Of course, as we are interested purely in the phenomenology of emotion we are concerned not with such 'action tendencies' themselves but with how they are manifested in experience. Thus, our interest is not in the fact that fear disposes us to flee but that, as Deonna and Teroni (2012) put it, in fear I have a sense of my body as *poised* to flee. Emotional experience, it seems, is partly constituted by the feeling of bodily expressions of emotion and also the experience of our bodies as *poised* or *ready* to act in ways appropriate to that emotion.

1.5 Cognitivism vs. non-cognitivism

The above – intentionality, feeling, evaluation, expression, and action – are among the features that theories of emotion typically attempt to accommodate. Although a simplification, we can think of such theories as, broadly speaking, falling into one of two groups. On the one hand we have cognitivist accounts that treat emotions as closely associated with belief or judgement and, on the other, we have non-cognitivist accounts that think of emotional experience as a more primitive phenomenon, largely independent of such higher faculties.

The non-cognitivist account that has had the greatest influence on contemporary philosophical discussion is that offered by James (1884). According to the standard interpretation of James's view, an emotion is the awareness, from the inside, of one's bodily reaction to some perceived event. This, as James recognises, reverses the ordinary order of explanation:

> Our natural way of thinking about these standard emotions is that the mental perception of some fact excites the mental affection called the emotion, and that this latter state of mind gives rise to the bodily expression. My thesis on the contrary is that *the bodily changes follow directly the PERCEPTION of the exciting fact and that the feeling of the same changes as they occur IS the emotion.*
>
> (1884, pp. 189–90)

On this view, emotional experience is a form of bodily awareness. Rather than the claim mentioned earlier that emotional experience is partly constituted by our awareness of how an emotion is expressed bodily, James thinks that the emotion is wholly constituted by our awareness of the body's reaction to some perceived object or event.

James's view has been criticised in a number of different ways by many philosophers (including Sartre, 1939b). One perennial criticism is that whilst the view might provide an acceptable account of the affective character of emotional experience, it seems particularly unsuited to accommodate either the intentionality of emotion or its connection with evaluation. If, as James suggests, emotions are feelings of bodily changes, then it would seem that the intentional object of emotion is, in every case, one's own body. But this flies in the face of the seeming truism that I am afraid of the ghost, not of my body; that I am proud of my children, not of my body. Nor, at least without further elaboration, does the account seem to have any place for evaluation. If fear is the feeling of my body's reaction to seeing a ghost, it is neither constituted by nor dependent on an evaluation of the ghost as a threat. We seem, on this view, to have lost sight of the formal objects of emotion.

Cognitive theories of emotion, on the other hand, focus precisely on intentionality and evaluation. The simplest cognitive views identify emotions with value judgements (see Nussbaum, 2001; Solomon, 2003). Thus, to fear the ghost *just is* to judge, and so evaluate, it as a threat. Such views have a number of benefits aside from their accommodation of these two features. For example, they make sense, in a way that the non-cognitivist seemingly cannot, of the intuitive thought that emotions are entered into for *reasons*. Thus, we ask *why* someone is afraid and criticise them if their answer does not provide an appropriate reason. This makes sense on the cognitivist view, since judgements are the sorts of things that we expect to be held for good reasons.

There are, however, problems with this simple cognitivist view. Given the discussion thus far, the most obvious is that the view looks unlikely to be able to account for the affective, felt character of emotion. Judgement, it might be argued, does not feel a certain way and, even if it does, it seems a poor candidate to capture the rich bodily phenomenology of emotional experience. Thus, it seems entirely possible to 'coldly' judge the apparition before me to be a threat, without in any way *feeling* afraid (Scheler, 1913/16, pp. 172–3; Goldie, 2000, Ch. 3). A second worry is that emotions often seem to be immune to the sorts of rational considerations that we ordinarily take to affect belief. Thus, to employ a familiar example, I may fear flying despite my acknowledging that it does not really constitute a threat to my well-being. That is, some emotions are recalcitrant. How, we might ask, can emotion be identified with judgement if it is, at least in some cases, immune to revision in light of evidence?

Cognitivists can respond to these concerns in a number of ways. They may, for example, seek to identify the intentional and feeling aspects of emotional experience, endorsing something like Goldie's (2000, Ch. 3) notion of 'feeling towards'. Alternatively, they may relinquish the identification of emotional experience with judgement, thinking instead of emotions as more akin to perceptual experience – as the *perception* of affective qualities (de Sousa, 1987). The extent to which these strategies succeed is a matter of dispute. I will not pursue the issue any further, rather turning to discussions of emotional phenomena located explicitly within the Phenomenological tradition. In particular, we will examine the views of both Heidegger and Sartre. I have discussed the above issues at such length for the reason that they help us to see what these two philosophers are claiming, and to evaluate the contribution that they can be thought to make to the contemporary discussion.

2 Sartre's Sketch

'Emotion', Sartre tells us in his *Sketch for a Theory of the Emotions*, 'is a mode of our conscious existence, one of the ways in which consciousness understands [...] its Being-in-the-World' (1939b, p. 91). We have encountered the Heideggerian concept of being-in-the-world in previous chapters. There are a number of aspects of this rich notion, discussion of which would take us too far from the current topic, but it is worth summarising a few of the central ideas. This will put us in a position to appreciate Sartre's account of the emotions as magical transformations of the world in the face of difficulty.

2.1 Sartrean being-in-the-world

Heidegger claims that in ordinary, everyday experience we are aware of entities as *equipment*, i.e. as entities that are *for* something (1927a, §15). That is, we experience entities in our environment in ways relevant to the particularities of our practical lives. I experience a pen as *for writing*, a chair as *for sitting on*, and so on, and such qualities are revealed not by way of any kind of theoretical investigation, but by a practical engagement with them:

> Equipment can genuinely show itself only in dealings cut to its own measure (hammering with a hammer, for example); but in such dealings an entity of this kind is not *grasped* thematically as an occurring Thing [...] the less we just stare at the hammer-Thing, and the more we seize hold of it and use it, the more primordial does our relationship to it become, and the more unveiledly is it encountered as that which it is – as equipment.
>
> (1927a, p. 98)

In Merleau-Ponty's words, entities are given as '[c]lothed in human qualities' (1948, p. 56), and engagement with them is an embodied activity, in which the body is 'the vehicle of being-in-the-world' (1945, p. 84), constituting an active, pre-objective openness that makes possible the experience of *things*. Sartre's discussion follows suit, claiming that everyday 'unreflective conduct' – the various bodily activities in which we are continuously engaged – involves an awareness of the world around us as a 'hodological chart' (1939b, p. 62). That is, as a series of paths leading to the various goals that we wish to accomplish:

> in a normal and well-adapted activity the objects 'to be realized' present themselves as needing to be realized in specific ways. The means themselves appear to us as potentialities that lay claim to existence. This apprehension of the means as the one possible path to the attainment of the end [...] may be called the pragmatic intuition of the determinism of the world. From this point of view, the world around us – what the Germans call the *Umwelt* – the world of our desires, our needs and of our activities, appears to be all furrowed with strait and narrow paths leading to such and such a determinate end.
>
> (1939b, p. 62)

According to Heidegger, Merleau-Ponty, and Sartre, then, the world around us is not given as populated by things given in the sorts of terms that would be recognised by the natural sciences, but is rather presented in terms that are significant to us. In short, worldly entities are presented as things about which we *care*. To employ Sartre's metaphor, the world appears as a series of routes that may be taken to the attainment of our goals. The 'determinism of the world' about which Sartre speaks here is the fact that such paths are experienced as constrained by basic causal facts: to get a drink, I must open the fridge door; to get to the café, I must turn left, and so on. This is crucial, since it is precisely this feature of our ordinary experience that, on Sartre's view, is suspended in emotional experience.

2.2 Difficulty and magic

If we experience the world around us as populated by paths to the attainment of our goals, then sometimes those paths are blocked or, at least, home to entities that may prevent us from travelling them. In Sartre's words, 'the world is *difficult*' (1939b, p. 62). It is the experience of difficulty, claims Sartre, that is the source of emotion which he thinks of as a transformation of the world in which we live:

> We can now conceive what an emotion is. It is a transformation of the world. When the paths before us become too difficult, or when we cannot see our way, we can no longer put up with such an exacting and difficult world. All ways are barred and nevertheless we must act. So then we try to change the world; that is, to live it as though the relations between things and their potentialities were not governed by deterministic processes but by magic.
>
> (1939b, p. 63)

The idea here is that emotion is a way of escaping from an intolerably difficult situation. When our path is in some way blocked, rather than seek a way to resolve the problem we 'transform' the world so as to nullify the causal facts. In emotional experience, on Sartre's view, rather than experience 'the determinism of the world', I am aware of the entities around me as though they obeyed magical laws. Since being-in-the-world involves being presented with things as replete with qualities that relate to my own practical interests, when my interests change, the way that the world appears can change. Thus, emotion is conceived as a response to an encountered difficulty: rather than solve the difficulty in the appropriate way through action, I 'solve' it by transforming my *awareness* of the world. Emotion is an

> unreflective consciousness which now grasps the world differently, under a new aspect, and imposes a new behaviour – through which that aspect is grasped – and this again serves as *hyle* for the new intention [...] during emotion, it is the body which, directed by the consciousness, changes its relationship with the world so that the world should change its qualities.
>
> (1939b, pp. 64–5)

This connection to bodily behaviour is important. According to Heidegger, the most fundamental way in which the characteristic qualities of some item of equipment are revealed to us is by using it. Similarly, Sartre claims that the way in which one transforms the world and comes to be aware of 'new aspects' (of which more below) is through bodily action. Sartre illustrates this account by sketching a number of cases. For example, concerning what he calls 'passive fear', Sartre writes:

> I see a ferocious beast coming towards me: my legs give way under me, my heart beats more feebly, I turn pale, fall down and faint away [...] it is a behaviour of *escape*; the fainting away is a refuge [...] unable to escape the danger by normal means and deterministic procedures, I have denied existence to it [...] I can suppress it as an object of consciousness, but only by suppressing consciousness itself.
>
> (1939b, p. 66)

The claim here is that in the face of an insurmountable difficulty, I act in such a way as to transform the world into one governed by magic. Of course, I do not really transform the world, I do not really annihilate the ferocious beast. But, in a world ruled by magic, such considerations have no place. I solve my problem, in a sense, by wishing it away.

It may seem as though, even if it is accepted for negative emotions such as fear, an account such as this will be implausible when applied to positively valenced emotions, such as joy. Sartre, however, does attempt an explanation in this case, telling us that '[j]oy is magical behaviour which tries, by incantation, to realize the possession of the desired object as an instantaneous totality' (1939b, p. 72). For example:

> a man to whom a woman has just said that she loves him may begin to dance and sing. In so doing he turns his mind away from the prudent and difficult behaviour he will have to maintain if he is to deserve this love [...]. For the moment, he is possessing the object by magic.
>
> (1939b, p. 73)

Sartre's account of emotion, both positive and negatively valenced, prefigures the description of bad faith that he later gives in *Being and Nothingness*. Sartre sees emotion as an essentially irrational phenomenon. When rational means fail me, I adopt an irrational approach, thereby coming to experience the world as one in which I can act by magic.

As mentioned above, Sartre thinks of emotional experience as involving the awareness of its objects as possessing aspects that they were not previously experienced as having. In passive fear, for example, the feared thing is experienced as possessing 'lesser presence' (1939b, p. 65); in passive sadness, 'the universe is bleak; it is, of undifferentiated structure' (1939b, p. 69); and so on. That objects are experienced as possessing such qualities is due to the bodily action that serves as an 'incantation'. But, if such qualities are placed there by consciousness, they are not given as such. Rather, in emotional experience we take them for real, independent qualities of things. 'Real emotion', Sartre maintains, 'is accompanied by belief. The qualities 'willed' upon the objects are taken to be real' (1939b, pp. 75–6). Thus, in sadness I do not just experience the world as bleak, I take it to be bleak; in horror, I do not simply experience the object as horrifying, I believe it to be so; and so on. In Sartre's words:

> Consciousness does not limit itself to the projection of affective meanings upon the world around it; it *lives* the new world it has thereby constituted – lives it directly, commits itself to it, and suffers from the qualities that the concomitant behaviour has outlined.
>
> (1939b, p. 78)

Sartre, however, has a very particular conception of belief in mind, understanding it in a bodily way, in terms of the physiological processes that accompany the different emotions. In fear, I feel the hairs on my neck stand up, in joy I experience a certain bodily relaxation. It is by way of undergoing these bodily changes in response to the affective qualities I experience, claims Sartre, that I take them for real:

> Merely to run away from it would not be enough to constitute an object as horrifying [...] we must be spell-bound and filled to overflowing by our emotion [...]. Here we can understand the part played by the purely physiological phenomena; they represent the *genuineness* of the emotion, they are the phenomena of belief.
>
> (1939b, p. 76)

2.3 Sartre on emotional experience

How, we might ask, does Sartre's account of emotional experience relate to those features of emotional experience that we sketched in §1? How, that is, does Sartre's view relate to the apparently commonsense view of emotions as intentional experiences that involve evaluation, feel a certain way, and are associated with particular forms of bodily expression and action? Further, is Sartre's picture cognitivist, non-cognitivist, or something else entirely?

In keeping with his account of consciousness more generally, Sartre is explicit that emotional experience is intentional. 'It is obvious', he says, 'that the man who is frightened is afraid *of* something' (1939b, pp. 56–7). Indeed, the claim that emotional experience involves a transformation of the 'deterministic' world into a world of 'magic' clearly indicates that, on his view, '[e]motion is a specific manner of apprehending the world' (1939b, p. 57).

Closely related to the intentionality of emotional experience is its evaluative role. In fear, I evaluate the feared object as a threat, in joy I evaluate it as wonderful, and so on. Once more, Sartre explicitly endorses this idea. As should be clear from the above, Sartre thinks of emotional experience as involving 'the projection of affective meanings upon the world' (1939b, p. 78). In being sad, for example, I experience the world as *bleak*, in horror I experience the object as *horrible*. Such affective qualities are evidently evaluative, providing experienced objects with a significance for us.

Closely related to the evaluative character of emotional experience is the fact that emotions feel a certain way. For a plausible claim is that the valence of an emotion – its feeling good or bad – is aligned with the nature of the value with which it is associated. So, being horrified feels bad, at least in part, because of the fact that the affective quality that one experiences the object to have, *horribleness*, has negative value. Joy, by contrast, feels good, at least in part, for the reason that the affective quality that one experiences the object to have, *wonderfulness*, has positive value. I suggested in §1, however, that the felt valence of an emotion does not exhaust the way that it feels. Rather, emotions typically have a bodily phenomenology. To see how Sartre's account accommodates this thought, we must turn to the role of action and expression in emotional experience.

We can distinguish between the bodily changes, bodily expressions, and bodily actions associated with emotional experience. Thus, in fear the hairs on my neck may rise (bodily change), I may widen my eyes and adopt an aghast look (bodily expression), and be motivated to flee the feared object (bodily action). As suggested above, Sartre places the body at the centre of his account of emotion. He speaks of a bodily 'incantation' by means of which the world is transformed into one governed by magic. As he says, '[e]motional behaviour seeks by itself, and without modifying the structure of the object, to confer another quality upon it' (1939b, p. 65). It is somewhat unclear, from Sartre's discussion, whether this 'emotional behaviour' is primarily a matter of bodily change, expression, or action. We get a slightly clearer picture, however, when we consider how Sartre's position sits with respect to the debate between cognitivism and non-cognitivism.

2.4 Sartrean cognitivism

If the crucial dividing line between cognitivists and non-cognitivists concerns whether or not emotional experience is conceived of as identical to, or otherwise dependent on, belief, then surely Sartre's account is a form of cognitivism. Commenting on the escapist nature of emotion, he writes that: '[i]f emotion is play-acting, the play is one that we believe in' (1939b, p. 65). 'Real emotion', he says, 'is accompanied by belief' (1939b, p. 75). As we saw, however, what Sartre means by belief in this context is far from the purely cognitive state that we tend to have in mind when we employ that term. Belief, on Sartre's account, is a matter of 'living' the magical world, of *committing* oneself to it, and this is constituted by the bodily changes – racing heart, raised hairs, etc. – mentioned above. We can, perhaps paradoxically, think of Sartre as offering a non-cognitive construal of belief in much the same way that Solomon (2003) – himself heavily indebted to Sartre – does.

This tells us a number of things. First, this construal of belief provides Sartre with a potential line of defence against a number of objections that can be raised against cognitivist views. For the fact that Sartre thinks of belief in terms of 'bodily disturbance' suggests both that his is a cognitivist account that does not ignore the felt aspect of emotion, and that he may have some immunity to worries about recalcitrant emotions. The former point is reasonably clear, since bodily disturbances *feel* some way or other; the latter is due to the fact that bodily disturbances are not the sort of thing that we ordinarily think of as affected by explicit conscious reasoning. As Sartre points out, 'one can stop oneself from running but not from trembling' (1939b, p. 77). To return to the fear of flying example mentioned earlier, it may be that whilst I accept the evidence-based proposition that flying is not dangerous, my body *lives* the situation as though it were capable of threatening me in a magical way. Of course, much more would need to be said on this score in defence of Sartre's position.

Second, we earlier left the question as to whether the 'emotional behaviour' that serves as an 'incantation' of the magical world is constituted primarily by bodily changes, bodily expressions, or bodily actions. Given what we have said so far, it seems plausible to suppose that Sartre is distinguishing between bodily changes, on the one hand, and bodily expression and action, on the other. Taking the example of terror of some object, bodily expressions and actions, on Sartre's view, seem to be what 'confer the formal quality "horrible" upon it'. Purely physiological bodily changes, on the other hand, are what fill it 'with something opaque and weighty that gives it substance' (1939b, p. 76). In short, bodily expression and action confer the affective quality on the object; bodily changes constitute one's 'believing' in it.

All of the above indicates that whilst there is a clear sense in which it is appropriate to think of Sartre's account as a form of cognitivism, the account nevertheless possesses certain aspects more commonly associated with noncognitivism. For, despite the significant differences between Sartre's and James's accounts of emotion, there does seem to be a similarity. Recall that, on James's view, emotion follows from and does not itself cause the bodily behaviour with which it is associated. Now that we have Sartre's view before us, we can see that something similar can be said of it. That is, the bodily action that transforms the world, and the bodily changes that sustain a commitment to that world, do not result from the emotional experience but rather feed into it. Thus, like James, Sartre seems to have reversed the ordinary conception of emotion as the source of emotional expression and action.

Notably, this is the exact opposite of the description of emotional expression with which Stein provides us. In Chapter 8, I mentioned the Steinian claim that we experience emotions as 'pouring into' bodily expression. Thus, on that view, reflection upon emotional experience teaches us that we weep because we are sad. On the view with which Sartre presents us, on the other hand, it seems that we should rather say that we are sad because we weep.

In fact, just as we saw in relation to his account of reflection and the transcendent ego, Sartre seems content to overturn such commonsense conceptions of the causal order of emotional experience. For example, according to our ordinary way of conceiving of emotion, we would see 'consciousness only as it is motivated by its object: "I am angry because *he* is hateful"'

(1939b, p. 91). This, however, is precisely the opposite of the truth according to Sartre, who claims that we can 'perceive emotion at work constituting the magical form of the world. "I find him hateful *because* I am angry"' (1939b, p. 91). Sartre is, thus, willing to turn common sense on its head when considering the relation between emotion and the awareness of affective qualities. The same seems to be true with respect to emotion and bodily expression. That is, whilst we typically think that we smile because we feel joy, Sartre apparently endorses the Jamesian reversal, supposing that, at least in part, we are happy because we smile. He would reject, then, the Steinian description of emotional expression.

2.5 Another kind of magic

These reversals of common sense are liable to seem implausible to many. The claim, for example, that we experience affective qualities such as dangerousness *because* we are afraid, rather than the other way around, may seem not to merit serious consideration. Surely, it will be said, fear is a *response* to our awareness of danger. Further, the claim that emotions are one and all evasive strategies in the face of intolerable difficulties is, for many, too much to swallow. Perhaps some emotional experiences have this sort of self-deceptive character, but surely Sartre has done little to convince us that every emotional episode is of this sort.

There is some reason to think that even Sartre may have had such concerns since, in his own discussion, he presents an example that does not take this form. That is, it seemingly neither adheres to the reversal of common sense according to which we are aware of an affective quality because we make ourselves emotional, nor involves any sort of self-deceptive evasive strategy in the face of an experienced difficulty. Sartre introduces the case as follows:

> This theory of emotion does not explain the immediate reactions of horror and wonder that sometimes possess us when certain objects suddenly appear to us. For example, a grimacing face suddenly appears pressed against the outside of the window; I am frozen with terror. Here, of course, there is no appropriate behaviour and it would seem that the emotion has no finality.
>
> (1939b, p. 84)

On the face of it, this case does not fit Sartre's view at all but is close to the commonsense picture according to which emotional experience is a reaction to the awareness of some affective quality, in this case *horribleness*. The example does involve magic, but not magic that, as with those emotions with which Sartre has been primarily concerned, emanates from oneself but rather from the world. In Sartre's words, in such cases 'it is this world that reveals itself to consciousness as magical just where we expect it to be deterministic' (1939b, p. 84). The magic, in the case at hand, consists in the fact that

> That face [...] we do not at first take it as that of a man, who might push the door open and take thirty paces to where we are standing. On the contrary, it is presented, motionless though it is, as acting at a distance. The face [...] annihilates the distance and enters into us.
>
> (1939b, p. 86)

This seemingly commonplace example does retain a number of features of Sartre's general account. As we have seen, it involves the awareness of an affective quality and the experience is characterised as one in which the world is presented to us as governed by magic. Further, although he is less than explicit, it seems that Sartre also thinks of such cases as involving belief (in his characteristic sense), since he thinks that 'consciousness seizes on the magic as magic, and lives it vividly as such' (1939b, p. 86).

What we thus far lack, however, is any sense of what it is that allows, in cases of this sort, the object to appear as *horrible*. That is, we do not know how it can be that 'the world of the utilizable vanishes abruptly and the world of magic appears in its place' (1939b, pp. 90–1). In cases such as these, there is seemingly nothing to perform the role of bodily 'incantation' that *makes* the object appear as possessing magical qualities. It just does. Sartre does attempt to give such an explanation, suggesting that 'the behaviour which gives its meaning to the emotion is no longer *our* behaviour; it is the expression of the face and the movements of the body of the other being' (1939b, p. 87). But it is far from clear that he is right to claim that this is 'of the same structure' as his standard cases, since it merely seems to restate the commonplace observation that the face appears horrible due to its expression.

So magic, in cases of this sort, seems not to be explained in the same way as it is in those cases such as sadness and joy with which Sartre primarily concerns himself. Indeed, Richmond (2011, 2014) argues that the account is in fact inconsistent in the role assigned to magic in the other cases. Rather than pursue this question further, however, it will be worthwhile to turn to Heidegger's account of emotions and, in particular, moods. For, as Richmond (2011) also points out, there is some affinity between cases such as the horrible face at the window and the Heideggerian account of emotional phenomena more generally. In fact, in his account of mood it seems that Heidegger may provide something like an answer to the above question of how it could be that a face could spontaneously appear as horrible, despite the fact that 'the horrible is *not possible* in the deterministic world of tools' (1939b, p. 89).

3 Attunement

Sartre's account of emotion draws upon the Heideggerian notion of being-in-the-world. According to Sartre, emotional consciousness transforms the 'deterministic' world of tools into one governed by magic. Since Sartre thinks of emotional consciousness as a transformation of the 'deterministic' world, it is, for him, just one possible way we have of being-in-the-world. Plausibly, then, he allows for the possibility of an entirely unemotional form of being-in-the-world, one in which the tools of which we are aware offer no hint of such magic. This, however, is something with which Heidegger himself would take issue. In particular, Heidegger's account of moods, and their relation to emotion, is inconsistent with this implication of Sartre's position. For, on Heidegger's account, mood is not an occasional lapse, but a necessary feature of being-in-the-world. Given this, and since we have yet to explicitly discuss moods, it is worth briefly turning to the Heideggerian account. Doing so will help to illuminate the above discussion of Sartre.

Heidegger thinks of moods and emotions as ways of 'being attuned' to the world and, as such, takes them to be an essential aspect of being-in-the-world: 'Mood has always already disclosed being-in-the-world as a whole and first makes possible directing oneself towards something' (1927a, p. 133). Attunement is a 'basic existential species' of the disclosure of both oneself (more properly, for Heidegger, 'Dasein') and the world. It is in being attuned to things through mood and emotion that things *matter*, so can *be there for me*.

> This mattering to it [Dasein] is grounded in attunement, and as attunement it has disclosed the world, for example, as something by which it can be threatened. Only something which is the attunement of fearing, or fearlessness, can discover things at hand in the surrounding world as being threatening. The moodedness of attunement constitutes existentially the openness to world of Dasein.
>
> (1927a, pp. 133–4)

In the earlier discussion of the intuitive relation between emotion and evaluation, I suggested that we would ordinarily find it plausible to suppose that we experience emotions only in relation to things about which we care, about things that matter to us. At a first pass, we can understand this as the thought that emotions just are ways of caring about things, of allowing things to matter to one. In the above quotations, however, Heidegger outlines a more fundamental role for *mattering*. For he suggests, first, that in order to undergo an emotional experience towards some object, it must *already* matter to you and, second, that an object's mattering is a necessary condition of its *being there* for you at all. Furthermore, he claims, something's mattering to you just is a matter of your being attuned to it by way of being in a certain mood.

Consider again Sartre's account of the motivation to enter emotional experience. Upon finding the world difficult, I magically transform it *via* a bodily 'incantation'. But, on Sartre's view, something's appearing difficult depends on my already having an awareness of the relationship between the objects around me and the possibilities they afford for satisfying my goals. That is, objects must already be significant to me; they must matter. But according to Heidegger, being attuned to the world just is being-in-the-world in such a way that things matter for one, and matter in a particular way. This mattering in a particular way is mood. Thus, in cheerfulness one is attuned to the world in one way; in anxiety, or boredom, one is attuned to it quite differently. That is, in each of these moods, the significance of one's surroundings is affected. As such, a mood is like an *atmosphere* or a *vibe*. There is always an atmosphere of some sort, and it partly determines what is *there* for you since it partly determines how it is that your surroundings matter.

In Chapter 2 mood was mentioned as a potential counterexample to Brentano's Thesis that all and only mental phenomena are intentional. There I briefly sketched the common response that rather than lacking intentional objects, moods actually have very general objects. Given what I have just said about Heidegger's view of moods, we can see that this is a response that he would likely endorse. For a mood, on Heidegger's account, is a way of being attuned to the world and one's place in it, in short to being-in-the-world. As Heidegger says, for the case of one mood to which he accords special importance, '*That in the face of which one has anxiety is Being-in-the-world as such*' (1927a, p. 180).

Mood makes possible things' mattering to one. It therefore makes possible emotional experience, which seems to depend on such mattering. It is for this reason that Heidegger claims that '[o]nly something which is the attunement of fearing, or fearlessness, can discover things at hand in the surrounding world as being threatening' (1927a, p. 134). It is being in some mood that makes it possible for things to appear as possessing affective qualities, such as *being threatening*. If this is correct, then it arguably gives us a way to answer the question with which we left Sartre's account of emotional experience.

On Sartre's official view, we imbue things with affective qualities by an act of magic. Sartre recognised, however, that not all emotional experiences seem to fit this picture. Not all emotional experiences involve a 'bodily incantation'. On the contrary, it seems entirely commonplace to experience the world as possessing affective qualities – the face is horrible – in an immediate way, without any intermediary bodily act on my part. On the Heideggerian picture that I have been sketching, it is the fact that one is always in some mood – not some or other concrete act of 'bodily incantation' – that makes possible the appearance of affective qualities. The Heideggerian, therefore, would not see such cases of immediate emotional reaction as in any way problematic. If, like Heidegger and unlike Sartre, we see mood as an ineliminable aspect of all being-in-the-world, then we will have a picture that, in this respect at least, stays truer to our ordinary view of emotional experience; one that is truer, one may well suppose, to the phenomena. And that, after all, is our goal.

4 Conclusion

That there is a deep connection between emotional experience and the evaluation of (aspects of) the world around us is a commonplace observation. Describing exactly what that connection might be is less straightforward. Sartre's account of emotions as 'magical' and Heidegger's account of mood as our basic way of being attuned to the world present powerful, yet importantly different, pictures of what it is for something to matter to us, and how our emotional lives are intertwined with perception, evaluation, and bodily experience.

11 Conclusion

Phenomenology, taken in a narrow sense, concerns the varieties of experience and the ways in which experienced things appear. Our experiential life is a complex weave of different elements, incorporating the perceptual experience of things and their properties, of events unfolding over time, the imaginary experience of merely possible things, of ourselves, our bodies, of others, and the emotional reactions we have to all of this. Pulling these strands apart, accurately describing them, and determining how they fit together is one of the most significant and difficult philosophical tasks. The preceding chapters provide only an initial glimpse of the multi-faceted complexity of our experiential lives. There is an element of the philosophical tradition that sees our own experience as that about which we are most certain, about which we simply cannot be mistaken. Husserl, certainly, falls within this tradition. But it would be a mistake to conflate the possibility of such certain knowledge with the idea that describing experience is easy. That it is not, that it is in fact one of the hardest things that we can attempt, is something that I have tried to indicate in this book. The patient, careful, and unbiased description of experience is – in true philosophical spirit – an endeavour that each of us can undertake from the comfort of our own armchairs, for it concerns nothing more than a painstaking examination of our everyday life as we live it.

But I have also tried to introduce something of the Phenomenological tradition, from its pre-history in Hume, Kant, and Brentano; through the ground-breaking pure phenomenological investigations of Husserl, Stein, and others; the fundamental change of direction towards Heidegger's fundamental ontology and hermeneutic phenomenology; all the way to the synthesis of various Husserlian and Heideggerian concerns in the existential phenomenologies of Sartre and Merleau-Ponty. Each of these thinkers engages in phenomenology in the narrow sense but also presents a broader picture in which narrowly phenomenological issues can be placed within the context of metaphysics, epistemology, the philosophy of language, ethics, and more. I have, as much as is feasible, avoided this broader debate. Presenting an account of the ways in which even this reasonably small number of philosophers see the relation between their phenomenological discussions and this broader context is a task that I could not possibly hope to complete within a single book. As a result, the discussion and investigations undertaken in this book are only partial, offering a way in to phenomenology. But it is a concern with phenomenology in the narrow sense that unites philosophers as otherwise diverse as Husserl, Heidegger, and Sartre. It is, thus, the core on which one first must get a grip, before one can broaden one's vision. And it seems to me that the only way to get a real sense of phenomenology is to do it. It is for that reason that I have attempted to interrogate these thinkers, to test their claims in a way that is accessible to a contemporary philosophical readership. To appropriate a Heideggerian term, the most productive way to understand phenomenology is to *dwell* in it.

Suggested reading

General

In addition to the works listed in the bibliography, the following are recommended as general resources for understanding phenomenology.

Bernet, Rudolf, Iso Kern and Eduard Marbach. 1993. *An Introduction to Husserlian Phenomenology*. Evanston, IL: Northwestern.

Cerbone, David. 2006. *Understanding Phenomenology*. Chesham: Acumen.

Dreyfus, Hubert and Mark Wrathall, eds. 2006. *A Companion to Phenomenology and Existentialism*. Oxford: Blackwell.

Gallagher, Shaun and Dan Zahavi. 2012. *The Phenomenological Mind*, 2nd edition. London: Routledge.

Hammond, Michael, Jane Howarth and Russell Keat. 1991. *Understanding Phenomenology*. Oxford: Blackwell.

Luft, Sebastian and Søren Overgaard, eds. 2012. *The Routledge Companion to Phenomenology*. London: Routledge.

Moran, Dermot and Tim Mooney, eds. 2002. *The Phenomenology Reader*. London: Routledge.

Smith, David Woodruff and Amie Thomasson, eds. 2005. *Phenomenology and Philosophy of Mind*. Oxford: Clarendon.

Sokolowski, Robert. 2000. *Introduction to Phenomenology*. Cambridge: Cambridge University Press.

Primary reading list

The most important primary text for each chapter.

Chapter 1

Husserl, Edmund. 1917. Pure Phenomenology, its Method, and its Field of Investigation. In Dermot Moran and Timothy Mooney, eds. *The Phenomenology Reader*. London: Routledge, 2002, pp. 124–33.

Chapter 2

Brentano, Franz. 1874. *Psychology from an Empirical Standpoint*, Book 2, Ch. 1: 'The Distinction Between Mental and Physical Phenomena'. Edited by Oskar Kraus, English edition edited by Linda McAlister. London: Routledge, 1995.

Chapter 3

Heidegger, Martin. 1925. *History of the Concept of Time: Prolegomena*, §5: 'Intentionality'. Translated by Theodore Kisiel. Bloomington: Indiana University Press, 1985.

Chapter 4

Merleau-Ponty, Maurice. 1945. *Phenomenology of Perception*, Part 2, Ch. 3, §A: 'Perceptual Constants'. Translated by Donald Landes. London: Routledge, 2012.

Chapter 5

Husserl, Edmund. 1905. *On the Phenomenology of the Consciousness of Internal Time (1893–1917)*, §§7–33: 'Analysis of the Consciousness of Time'. Translated by John Barnett Brough. Dordrecht: Kluwer, 1991.

Chapter 6

Sartre, Jean-Paul. 1940. *The Imaginary*, Part I, Ch. 1: 'Description'. Translated by Jonathan Webber. London: Routledge, 2004.

Chapter 7

Sartre, Jean-Paul. 1937. *The Transcendence of the Ego*, Part I: 'The I and the Me'. Translated by Andrew Brown. London: Routledge, 2004.

Chapter 8

Stein, Edith. 1917. *On the Problem of Empathy*, Chapter 3, §4: '"I" and the Living Body'. Translated by Waltrout Stein. The Hague: Martinus Nijhoff, 1970.

Chapter 9

Stein, Edith. 1917. *On the Problem of Empathy*, Chapter 3, §5: 'Transition to the Foreign Individual'. Translated by Waltrout Stein. The Hague: Martinus Nijhoff, 1970.

Chapter 10

Sartre, Jean-Paul. 1939. *Sketch for a Theory of the Emotions*, Ch. 3: 'Outline of a Phenomenological Theory'. Translated by Philip Mairet. London: Routledge, 1981.

Further reading list

A selection of useful texts to consult in addition to those listed in the bibliography.

Chapter 1: The science of experience

Carman, Taylor. 2006. The Principle of Phenomenology. In Charles Guignon, ed. *Cambridge Companion to Heidegger*, 2nd edition. Cambridge: Cambridge University Press, pp. 97–119.

Casey, Edward. 1977. Imagination and Phenomenological Method. In Frederick Elliston and Peter McCormick, eds. *Husserl: Expositions and Appraisals.* Notre Dame, IN: University of Notre Dame Press, pp. 70–83.

Haaparanta, Leila. 2009. The Method of Analysis and the Idea of Pure Philosophy in Husserl's Transcendental Phenomenology. In Michael Beaney, ed. *The Analytic Turn: Analysis in Early Analytic Philosophy and Phenomenology.* London: Routledge, pp. 257–69.

Heidegger, Martin. 1925. *History of the Concept of Time: Prolegomena,* §§10–11. Translated by Theodore Kisiel. Bloomington: Indiana University Press, 1985.

Overgaard, Søren. 2002. Epoché and Solipsistic Reduction. *Husserl Studies* 18:3, pp. 209–22.

Chapter 2: The objects of experience

Alweiss, Lilian. 2009. Between Internalism and Externalism: Husserl's Account of Intentionality. *Inquiry* 52:1, pp. 53–78.

Dreyfus, Hubert and Harrison Hall. 1982. Introduction. In H. Dreyfus and H. Hall, eds. *Husserl, Intentionality, and Cognitive Science.* Cambridge, MA: MIT Press, pp. 1–27.

Jacquette, Dale. 2004. Brentano's Concept of Intentionality. In D. Jacquette, ed. *The Cambridge Companion to Brentano.* Cambridge: Cambridge University Press, pp. 98–130.

McAlister, Linda. 1974. Chisholm and Brentano on Intentionality. *Review of Metaphysics* 28:2, pp. 328–38.

Moran, Dermot. 1996. Brentano's Thesis. *Proceedings of the Aristotelian Society, Supplementary Volume* 70:1, pp. 1–27.

Chapter 3: Experiencing things

Cunningham, Suzanne. 1985. Perceptual Meaning and Husserl. *Philosophy and Phenomenological Research* 45:4, pp. 553–66.

Kind, Amy. 2007. Restrictions on representationalism. *Philosophical Studies* 134:3, pp. 405–27.

Macpherson, Fiona. 2006. Ambiguous Figures and the Content of Experience. *Noûs* 40:1, pp. 82–117.

Mooney, Tim. 2010. Understanding and Simple Seeing in Husserl. *Husserl Studies* 26:1, pp. 19–48.

Shim, Michael. 2011. Representationalism and Husserlian Phenomenology. *Husserl Studies* 27:3, pp. 197–215.

Chapter 4: Experiencing properties

Noë, Alva. 2005. Real Presence. *Philosophical Topics* 33:1, pp. 235–64.

Overgaard, Søren. 2010. On the Looks of Things. *Pacific Philosophical Quarterly* 91:2, pp. 260–84.

Romdenh-Romluc, Komarine. 2011. *Merleau-Ponty and the Phenomenology of Perception.* London: Routledge, Ch. 4.

Schellenberg, Susanna. 2008. The Situation-Dependency of Perception. *Journal of Philosophy* 105:2, pp. 55–84.

Siewert, Charles. 2006. Is Shape Appearance Protean? *Psyche* 12:3, pp. 1–16.

Chapter 5: Experiencing events

Anderson, Holly. 2014. The Development of the 'Specious Present' and James' views on Temporal Experience. In Dan Lloyd and Valtteri Arstilla, eds. *Subjective Time*. Cambridge, MA: MIT Press, pp. 25–42.

Dainton, Barry. 2008. Sensing Change. *Philosophical Issues* 18:1, pp. 362–84.

Hoerl, Christoph. 2009. Time and Tense in Perceptual Experience. *Philosophers' Imprint* 9:12, pp. 1–18.

Rodemeyer, Lanei. 2006. *Intersubjective Temporality: It's About Time*. Dordrecht: Springer.

Zahavi, Dan. 2007. Perception of Duration Presupposes Duration of Perception – or Does it? Husserl and Dainton on Time. *International Journal of Philosophical Studies* 15:3, pp. 453–71.

Chapter 6: Experiencing possibilities

De Preester, Helena. 2012. The Sensory Component of Imagination: The Motor Theory of Imagination as a Present Day Solution to Sartre's Critique. *Philosophical Psychology* 25:4, pp. 503–20.

Jansen, Julia. 2005. On the Development of Husserl's Transcendental Phenomenology of Imagination and its Use in Interdisciplinary Research. *Phenomenology and the Cognitive Sciences* 4:2, pp. 141–32.

Ricoeur, Paul. 1981. Sartre and Ryle on Imagination. In Paul Schilpp, ed. *The Philosophy of Jean-Paul Sartre*. La Salle, IL: Open Court, pp. 167–79.

Stawarska, Beata. 2005. Defining Imagination: Sartre between Husserl and Janet. *Phenomenology and the Cognitive Sciences* 4:2, pp. 133–53.

Tye, Michael. 1991. *The Imagery Debate*. Cambridge, MA: MIT Press, Chs 1–2.

Chapter 7: Experiencing oneself

Carr, David. 1999. *The Paradox of Subjectivity: The Self in the Transcendental Tradition*. New York: Oxford University Press, Ch. 3.

Stawarska, Beata. 2002. Memory and Subjectivity: Sartre in Dialogue with Husserl. *Sartre Studies International* 8:2, pp. 94–111.

Sutton Morris, Phyllis. 1985. Sartre on the Transcendence of the Ego. *Philosophy and Phenomenological Research* 46:2, pp. 179–98.

Wider, Kathleen. 1997. *The Bodily Nature of Consciousness: Sartre and Contemporary Philosophy of Mind*. Ithaca, NY: Cornell University Press, Ch. 3.

Zahavi, Dan. 2000. Self and Consciousness. In D. Zahavi, ed. *Exploring the Self*. Amsterdam: John Benjamins, pp. 55–74.

Chapter 8: Experiencing embodiment

Gallagher, Shaun. 1986. Lived Body and Environment. *Research in Phenomenology* 16:1, pp. 139–70.

Heinämaa, Sara. 2012. The Body. In Sebastian Luft and Søren Overgaard, eds. *The Routledge Companion to Phenomenology*. London: Routledge, pp. 222–32.

Kelly, Sean. 2002. Merleau-Ponty on the Body. *Ratio (New Series)* 15:4, pp. 376–91.

Romdenh-Romluc, Komarine. 2011. *Merleau-Ponty and the Phenomenology of Perception*. London: Routledge, Ch. 3.

Wider, Kathleen. 1997. *The Bodily Nature of Consciousness: Sartre and Contemporary Philosophy of Mind*. Ithaca, NY: Cornell University Press, Ch. 5.

Chapter 9: Experiencing others

Carr, David. 1973. The 'Fifth Meditation' and Husserl's Cartesianism. *Philosophy and Phenomenological Research* 34:1, pp. 14–35.

Dullstein, Monika. 2013. Direct Perception and Simulation: Stein's Account of Empathy. *Review of Philosophy and Psychology* 4:2, pp. 333–50.

Gallagher, Shaun. 2005. Phenomenological Contributions to a Theory of Social Cognition. *Husserl Studies* 21:2, pp. 95–110.

Staehler, Tanja. 2008. What Is the Question to Which Husserl's Fifth *Cartesian Meditation* Is the Answer? *Husserl Studies* 24:2, pp. 99–117.

Zahavi, Dan. 2014. Empathy and Other-Directed Intentionality. *Topoi* 33:1, pp. 129–42.

Chapter 10: Experiencing emotion

Cabestan, Philippe. 2004. What Is it to Move Oneself Emotionally? Emotion and Affectivity According to Jean-Paul Sartre. *Phenomenology and the Cognitive Sciences* 3:1, pp. 89–96.

Fell, Joseph P. 1965. *Emotion in the Thought of Sartre*. New York: Columbia University Press.

Hatzimoysis, Anthony. 2009. Emotions in Heidegger and Sartre. In Peter Goldie, ed. *The Oxford Handbook of Philosophy of Emotion*. Oxford: Oxford University Press, pp. 215–36.

Mendelovici, Angela. 2014. Pure Intentionalism About Moods and Emotions. In Uriah Kriegel, ed. *Current Controversies in Philosophy of Mind*. London: Routledge, pp. 135–57.

Tappolet, Christine. 2012. Emotions, Perceptions, and Emotional Illusions. In Calabi Clotilde, ed. *Perceptual Illusions: Philosophical and Psychological Essays*. Basingstoke: Palgrave-Macmillan, pp. 207–24.

Glossary

Each of the concepts described in this glossary is contested. Consequently, these descriptions should be taken as illustrative rather than definitive.

Adumbration The way in which something's features are given. A thing's constant shape, for example, is given in varying adumbrations as one moves around it.

Affiliation (also, the sense of ownership) The sense of one's body as one's own.

Analogising Transfer The transfer of the sense of some primal instituting to a new case.

Animation (also, construing) The transformation of perceptual sensations, by an intentional matter, into perceptual experiences of things.

Attunement An aspect of being-in-the-world. Through attunement things matter for one thereby making possible emotional experience towards them.

Being-in-the-World Dasein's way of being, a basic form of transcendence in which we are practically oriented towards a world structured in terms of our concerns, goals, and projects. On Heidegger's account, being-in-the-world is what makes intentionality possible.

Brentano's Thesis (also, Intentionality as the Mark of the Mental) The claim that all and only mental phenomena exhibit intentionality.

Dasein Heidegger's term for us, with fewer controversial connotations than the traditional 'subject'.

Ego, Empirical Oneself as an object of consciousness, revealed through reflection.

Ego, Pure Oneself as subject of consciousness, the 'subject pole' that Husserl claims is implicitly given in all intentional experience.

Eidetic Intuition The experience of the essence of a phenomenon.

Empathy (also, fellow-feeling) Entering into another's perspective, enabling us to know what things are like for them.

Environmental Thing (also, equipment) An experienced thing given in terms of its practical relevance to one's concerns, goals, and projects.

Fusion The identification of one's lived body with one's body given through outer perception.

Givenness   (also, presence) How intentional objects are for an experiencer. Things, for example, are given (or presented) as shaped, coloured, etc.

Givenness, Absolute   Given in such a way that all aspects of it are available to reflection, with nothing obscured. For example, sensations are absolutely given.

Givenness, Bodily   (also, bodily presence; in the flesh) Something's being given as there in person, as is the case with perceived things. According to Husserl, bodily givenness involves being given *via* hyle.

Giveness, Co-   (also, co-presented; appresented) Given as accompanying something bodily given but without itself being bodily given. The rear side of a perceived thing is co-given.

Hallucination   A perceptual experience in which one appears to be aware of something that does not in fact exist.

Horizon, Inner   Those aspects of a perceived thing that are merely co-presented.

Horizon, Outer   Those aspects of the surroundings of a perceived thing that are co-presented.

Hyle   (also, hyletic data; perceptual sensation; content; qualia) Non-intentional features of experience that are animated by intentions thereby coming to present objects.

Illusion   A perceptual experience in which one is aware of something that appears to be some way that in fact it is not.

Imagination, Sensory   Imagining some thing, a paradigm example of which is visualising.

Imagination, Suppositional   Imagining that something is the case.

Imagination, Variation in   Husserl's method of determining the essence of phenomena by finding the invariant through imagined variations on an experience.

Immanent   Contained within or a part of an intentional experience. Contrasted with Transcendent.

Intentionality   (also, representation) A non-relational directedness towards objects.

Intentional Inexistence   Brentano's characterisation of intentionality, often taken to involve a generalisation of the sense-data account of perception.

Intentional Matter   (also, meaning; representational content; interpretive sense) That aspect of an intentional experience in virtue of which it picks out a particular object and presents it as being some way.

Intentional Object   The object of an intentional state, what an intentional state purports to be about.

Intentional Quality   That aspect of an intentional experience in virtue of which it is the kind of experience it is, e.g. desire, visual perception, memory, etc.

Lived Body   (also, *Leib*) One's body as given from the inside as a sensitive body, at the centre of experiential space, and immediately responsive to the will.

Myth of the Given The claim that brute, non-conceptual sensory input cannot justify one conceptualisation over another.

Noema The intended as intended, variously interpreted as the sense of an intentional experience (cf. Intentional Matter) or the intentional object in its mode of givenness.

Noesis Those features that are intrinsic to an intentional experience itself, making it the sort of experience it is (cf. Intentional Quality).

Object in the Pregnant Sense (also, Object) An object the reality of which transcends our awareness of it, i.e. it is mind-independent.

Pairing When two things of the same kind are experienced, each being co-given in terms of the other (cf. Analogising Transfer).

Perceptual Anticipation The experience of something as fulfilling a conditional. For example, the experience of a thing as looking a certain way were it rotated (cf. Protention).

Perceptual Constancy Some thing's properties appearing to remain constant despite differences in the way that it seems with respect to that property (cf. Adumbration).

Phenomena (also, appearance) That which is given in an experience.

Phenomenal Principle The claim that the object of an intentional experience must have the properties it seems to have.

Positing The feature of intentional experiences that 'asserts' their actuality, or their absence, etc.

Pre-Ontological Understanding of Being The implicit grasp of what it is for something of a given sort to be, rather than not be.

Pre-Reflective Consciousness Experience that is not reflected upon.

Primal Impression The intentional awareness of the present phase of an event as *now happening*.

Primal Instituting The first instance of seeing something as being of a certain kind.

Protention The intentional awareness of a future phase of an event as *about to happen*.

Reduction, Eidetic The exclusion of contingent matters leading to an exclusive focus on essential features of phenomena.

Reduction, Phenomenological (also, bracketing; *epoché*) The exclusion of transcendent entities leading to an exclusive focus on that which is absolutely given.

Reflection The turning of one's attention to one's own experiences.

Retention The intentional awareness of a past phase of an event as *having just happened*.

Seeing-in The experience of one entity in virtue of the perception of another. Paradigmatically, the experience of a depicted object *via* the perception of a picture.

Sense-Data (also, impression; idea; image) Subjective entities sometimes claimed to be the objects of perceptual experience.

Specious Present   An extended period of time surrounding the actual present which it is sometimes claimed is, in some sense, given as present.

Transcendent   Not contained within or a part of an intentional experience. Contrasted with Immanent.

Transcendental   Necessary for the possibility of experience.

World   (also, environment) Our experienced surroundings. According to Heidegger the world is given as a context of meaningful relations between equipment and Dasein.

Bibliography

Armstrong, David. 1962. *Bodily Sensations*. London: Routledge.

Avramides, Anita. 2001. *Other Minds*. London: Routledge.

Bartok, Philip. 2005. Brentano's Intentionality Thesis: Beyond the Analytic and Phenomenological Readings. *Journal of the History of Philosophy* 43:4, pp. 437–460.

Bell, David. 1990. *Husserl*. London: Routledge.

Bermúdez, José Luis. 1998. *The Paradox of Self-Consciousness*. Cambridge, MA: MIT Press.

Bermúdez, José Luis. 2011. Bodily Awareness and Self-Consciousness. In Shaun Gallagher, ed. *The Oxford Handbook of the Self*. Oxford: Oxford University Press, pp. 157–179.

Blattner, William. 1999a. *Heidegger's Temporal Idealism*. Cambridge: Cambridge University Press.

Blattner, William. 1999b. Is Heidegger a Representationalist? *Philosophical Topics* 27:2, pp. 179–204.

Boghossian, Paul. 1997. What the Externalist Can Know *A Priori*. *Proceedings of the Aristotelian Society* 97:2, pp. 171–175.

Brentano, Franz. 1874. *Psychology from an Empirical Standpoint*. Edited by Oskar Kraus, English edition edited by Linda McAlister. London: Routledge, 1995.

Brentano, Franz. 1890–1. *Descriptive Psychology*. Translated and edited by Benito Müller. London: Routledge, 1995.

Brewer, Bill. 1995. Bodily Awareness and the Self. In José Luis Bermúdez, Anthony Marcel and Naomi Eilan, eds. *The Body and the Self*. Cambridge, MA: MIT Press, pp. 291–309.

Brewer, Bill. 2011. *Perception and its Objects*. Oxford: Oxford University Press.

Carman, Taylor. 1994. On Being Social: A Reply to Olafson. *Inquiry* 37:2, pp. 203–223.

Carman, Taylor. 1999. The Body in Husserl and Merleau-Ponty. *Philosophical Topics* 27:2, pp. 205–226.

Carman, Taylor. 2003. *Heidegger's Analytic: Interpretation, Discourse, and Authenticity in Being and Time*. Cambridge: Cambridge University Press.

Casey, Edward. 2000. *Imagining: A Phenomenological Study*. Bloomington: Indiana University Press.

Christensen, Carleton B. 1997. Heidegger's Representationalism. *Review of Metaphysics* 51:1, pp. 77–103.

Christensen, Carleton B. 1998. Getting Heidegger off the West Coast. *Inquiry* 41:1, pp. 45–68.

Cole, Jonathan. 1995. *Pride and a Daily Marathon*. Cambridge, MA: MIT Press.

Coplan, Amy and Peter Goldie. 2011. *Empathy: Philosophical and Psychological Perspectives*. Oxford: Oxford University Press.

Crane, Tim. 1998. Intentionality as the Mark of the Mental. In Anthony O'Hear, ed. *Contemporary Issues in the Philosophy of Mind*. Cambridge: Cambridge University Press, pp. 229–252.

Crane, Tim. 2006a. Brentano's Concept of Intentional Inexistence. In Mark Textor, ed. *The Austrian Contribution to Philosophy*. London: Routledge, pp. 20–35.

Crane, Tim. 2006b. Is There a Perceptual Relation? In Tamar Gendler and John Hawthorne, eds. *Perceptual Experience*. Oxford: Oxford University Press, pp. 126–146.

Crowell, Steven. 2001. *Husserl, Heidegger, and the Space of Meaning: Paths Toward Transcendental Phenomenology*. Evanston, IL: Northwestern University Press.

Crowell, Steven. 2004. Heidegger and Husserl: The Matter and Method of Philosophy. In Hubert Dreyfus and Mark Wrathall, eds. *Blackwell Companion to Heidegger*. Oxford: Blackwell, pp. 49–64.

Dainton, Barry. 2000. *Stream of Consciousness*. London: Routledge.

Dainton, Barry. 2010. Temporal Consciousness. In Edward N. Zalta, ed. *The Stanford Encyclopedia of Philosophy* (Spring 2014 edition).

de Beauvoir, Simone. 1947. *The Ethics of Ambiguity*. Translated by Bernard Frechtman. New York: Citadel Press, 1994.

de Beauvoir, Simone. 1960. *The Prime of Life*. Translated by Peter Green. New York: Harper & Row, 1962.

de Sousa, Ronald. 1987. *The Rationality of Emotion*. Cambridge, MA: MIT Press.

de Vignemont, Frédérique. 2007. Habeas Corpus: The Sense of Ownership of One's Own Body. *Mind and Language* 22:4, pp. 427–449.

Deonna, Julien and Fabrice Teroni. 2012. *The Emotions: A Philosophical Introduction*. London: Routledge.

Derrida, Jacques. 1967. *Speech and Phenomena*. In *Speech and Phenomena and Other Essays on Husserl's Theory of Signs*. Translated by David Allison. Evanston, IL: Northwestern University Press, pp. 1–104.

Descartes, René. 1641. *Meditations on First Philosophy*. In *Descartes: Selected Philosophical Writings*. Translated by John Cottingham, Robert Stoothoff, and Dugald Murdoch. Cambridge: Cambridge University Press, 1988, pp. 73–159.

Dreyfus, Hubert. 1991. *Being-in-the-World: A Commentary on Heidegger's Being and Time, Division 1*. Cambridge, MA: MIT Press.

Dreyfus, Hubert. 1993. Heidegger's Critique of the Husserl/Searle Account of Intentionality. *Social Research* 60:1, pp. 17–38.

Dreyfus, Hubert. 2005. Overcoming the Myth of the Mental. *Proceedings and Addresses of the American Philosophical Association* 79, pp. 47–65.

Dreyfus, Hubert. 2013. The Myth of the Pervasiveness of the Mental. In Joseph Schear, ed. *Mind, Reason, and Being-in-the-World*. London: Routledge, pp. 15–40.

Dummett, Michael. 1993. *Origins of Analytical Philosophy*. London: Duckworth.

Evans, Gareth. 1982. *The Varieties of Reference*. Edited by John McDowell. Oxford: Clarendon Press.

Fish, William. 2009. *Perception, Hallucination, and Illusion*. Oxford: Oxford University Press.

Fodor, Jerry. 1987. *Psychosemantics*. Cambridge, MA: MIT Press.

Føllesdal, Dagfinn. 1969. Husserl's Notion of Noema. *Journal of Philosophy* 66:20, pp. 680–687.

Føllesdal, Dagfinn. 2000. Absorbed Coping, Husserl and Heidegger. In Mark Wrathall and Jeff Malpas, eds. *Heidegger, Authenticity, and Modernity: Essays in Honor of Hubert L. Dreyfus, Volume 1*. Cambridge, MA: MIT Press, pp. 251–257.

Foster, John. 2000. *The Nature of Perception*. Oxford: Oxford University Press.

Frege, Gottlob. 1892. On *Sinn* and *Bedeutung*. Translated by Max Black. In Michael Beaney, ed. *The Frege Reader*. Oxford: Blackwell, 1997, pp. 151–171.

Frege, Gottlob. 1897. Logic. Translated by Peter Long and Roger White. In Michael Beaney, ed. *The Frege Reader*. Oxford: Blackwell, 1997, pp. 227–250.

Frege, Gottlob. 1918. Thought. Translated by Peter Geach and R. H. Stoothoff. In Michael Beaney, ed. *The Frege Reader*. Oxford: Blackwell, 1997, pp. 325–345.

Gallagher, Shaun. 2008. Direct Perception in the Intersubjective Context. *Consciousness and Cognition* 17:2, pp. 535–543.

Gardner, Sebastian. 2009. *Sartre's Being and Nothingness*. London: Continuum.

Gardner, Sebastian. 2015. Merleau-Ponty's Transcendental Theory of Perception. In Sebastian Gardner and Matt Grist, eds. *The Transcendental Turn*. Oxford: Oxford University Press, pp. 294–323.

Gennaro, Rocco. 2002. Jean-Paul Sartre and the HOT Theory of Consciousness. *Canadian Journal of Philosophy* 32:3, pp. 293–330.

Gettier, Edmund. 1963. Is Justified True Belief Knowledge? *Analysis* 23:6, pp. 121–123.

Gibson, James. 1979. *The Ecological Approach to Visual Perception*. Hillsdale, NJ: Lawrence Erlbaum.

Goldie, Peter. 2000. *The Emotions: A Philosophical Exploration*. Oxford: Oxford University Press.

Green, Mitchell. 2007. *Self-Expression*. Oxford: Clarendon Press.

Gunther, York. 2003. *Essays on Non-Conceptual Content*. Cambridge, MA: MIT Press.

Hall, Harrison. 1980. The Other Minds Problem in Early Heidegger. *Human Studies* 3:1, pp. 247–254.

Hannay, Alastair. 1971. *Mental Images: A Defence*. London: Unwin.

Heidegger, Martin. 1925. *History of the Concept of Time: Prolegomena*. Translated by Theodore Kisiel. Bloomington: Indiana University Press, 1985.

Heidegger, Martin. 1927a. *Being and Time*. Translated by Joan Stambaugh. Albany: State University of New York Press, 2010.

Heidegger, Martin. 1927b. *The Basic Problems of Phenomenology*. Revised edition, translated by Albert Hofstadter. Bloomington: Indiana University Press, 1982.

Hopkins, Robert. 1998. *Picture, Image and Experience*. Cambridge: Cambridge University Press.

Hopp, Walter. 2008. Husserl on Sensation, Perception and Interpretation. *Canadian Journal of Philosophy* 38:2, pp. 219–246.

Hopp, Walter. 2011. *Perception and Knowledge: A Phenomenological Account*. Cambridge: Cambridge University Press.

Howell, Robert J. 2010. Subjectivity and the Elusiveness of the Self. *Canadian Journal of Philosophy* 40:3, pp. 459–483.

Hume, David. 1739–40. *A Treatise of Human Nature*. Edited by L. S. Selby-Bigge, second edition revised by P. H. Nidditch. Oxford: Clarendon, 1978.

Husserl, Edmund. 1900–1. *Logical Investigations*. Translated by J. N. Findlay, edited by Dermot Moran. London: Routledge, 2001.

Husserl, Edmund. 1904–5. Phantasy and Image Consciousness. Translated by John Brough. In Husserl, E. *Phantasy, Image Consciousness, and Memory (1898–1925)*. Dordrecht: Kluwer, 2005, pp. 1–151.

Husserl, Edmund. 1905. *On the Phenomenology of the Consciousness of Internal Time (1893–1917)*. Translated by John Barnett Brough. Dordrecht: Kluwer, 1991.

Husserl, Edmund. 1907. *Thing and Space: Lectures of 1907*. Translated by Richard Rojcewicz. Dordrecht: Kluwer, 2010.

Husserl, Edmund. 1910–11. Philosophy as Rigorous Science. Translated and edited by Quentin Lauer, in *Philosophy and the Crisis of Philosophy*. New York: Harper & Row, 1965, pp. 71–148.

Husserl, Edmund. 1913. *Ideas Pertaining to a Pure Phenomenology and to a Phenomenological Philosophy, First Book*. Translated by F. Kersten. Dordrecht: Kluwer, 1998.

Husserl, Edmund. 1917. Pure Phenomenology, its Method, and its Field of Investigation. In Dermot Moran and Timothy Mooney, eds. *The Phenomenology Reader*. London: Routledge, 2002, pp. 124–133.

Husserl, Edmund. 1931. *Cartesian Meditations: An Introduction to Phenomenology*. Translated by Dorion Cairns. The Hague: Nijhoff, 1960.

Husserl, Edmund. 1952. *Ideas Pertaining to a Pure Phenomenology and to a Phenomenological Philosophy, Second Book*. Translated by Richard Rojcewicz and André Schuwer. Dordrecht: Kluwer, 1989.

Husserl, Edmund. 1954. *The Crisis of European Sciences and Transcendental Phenomenology*. Translated by David Carr. Evanston: Northwestern University Press, 1970.

Husserl, Edmund. 1997. *Psychological and Transcendental Phenomenology and the Confrontation with Heidegger (1927–1931)*. Translated and edited by Thomas Sheehan and Richard Palmer. Dordrecht: Kluwer.

Hyder, David and Hans-Jörg Rheinberger, eds. 2009. *Science and the Life-World: Essays on Husserl's Crisis of European Sciences*. Stanford, CA: Stanford.

Ingarden, Roman. 1931. *The Literary Work of Art: An Investigation of the Borderlines of Ontology, Logic, and the Theory of Language*. Translated by George Grabowicz. Evanston, IL: Northwestern University Press, 1973.

Jackson, Frank. 1977. *Perception*. Cambridge: Cambridge University Press.

James, William. 1884. What Is an Emotion? *Mind* 9, pp. 188–205.

James, William. 1890. *The Principles of Psychology*. New York: Holt & Co.

Kant, Immanuel. 1781/1787. *Critique of Pure Reason*. Translated by Paul Guyer and Allen W. Wood. Cambridge: Cambridge University Press, 1997.

Keller, Pierre. 1999. *Husserl and Heidegger on Human Experience*. Cambridge: Cambridge University Press.

Kelly, Sean. 1999. What Do We See (When We Do)? *Philosophical Topics* 27:2, pp. 107–128.

Kelly, Sean. 2000. Grasping at Straws: Motor Intentionality and the Cognitive Science of Skilled Behaviour. In Mark Wrathall and Jeff Malpas, eds. *Heidegger, Coping, and Cognitive Science: Essays in Honor of Hubert L. Dreyfus, Volume 2*. Cambridge, MA: MIT Press, pp. 161–177.

Kelly, Sean. 2005a. Seeing Things in Merleau-Ponty. In Taylor Carman and Mark Hansen, eds. *The Cambridge Companion to Merleau-Ponty*. Cambridge: Cambridge University Press, pp. 74–110.

Kelly, Sean. 2005b. The Puzzle of Temporal Awareness. In Andrew Brook and Kathleen Akins, eds. *Cognition and Neuroscience*. Cambridge: Cambridge University Press, pp. 208–238.

Kenny, Anthony. 1963. *Action, Emotion and Will*. London: Routledge & Kegan Paul.

Kind, Amy. 2001. Putting the Image Back in Imagination. *Phenomenology and the Cognitive Sciences* 62:1, pp. 85–110.

Levinas, Emmanuel. 1930. *The Theory of Intuition in Husserl's Phenomenology*. Second edition. Translated by André Orianne. Evanston, IL: Northwestern University Press, 1995.

Locke, John. 1690. *An Essay Concerning Human Understanding*. Edited by Peter Nidditch. Oxford: Clarendon Press, 1975.

Madary, Michael. 2010. Husserl on Perceptual Constancy. *European Journal of Philosophy* 20:1, pp. 145–165.

Martin, Michael. 1995. Bodily Awareness: A Sense of Ownership. In José Luis Bermúdez, Anthony Marcel, and Naomi Eilan, eds. *The Body and the Self*. Cambridge, MA: MIT Press, pp. 267–289.

McCulloch, Gregory. 1994. *Using Sartre: An Analytical Introduction to Early Sartrean Themes*. London: Routledge.

McDaniel, Kris. Forthcoming. Heidegger on the 'There Is' of Being. *Philosophy and Phenomenological Research*.

McDowell, John. 1982. Criteria, Defeasibility, and Knowledge. Reprinted in John McDowell, *Meaning, Knowledge, and Reality*. Cambridge, MA: Harvard University Press, 1998, pp. 369–394.

McDowell, John. 1994. *Mind and World*. Cambridge, MA: Harvard University Press.

McIntyre, Ronald. 1987. Husserl and Frege. *The Journal of Philosophy* 84:10, pp. 528–535.

McManus, Denis. 2012. *Heidegger and the Measure of Truth*. Oxford: Oxford University Press.

Meinong, Alexius. 1904. The Theory of Objects. Translated by Isaac Levi, D. B. Terrell and Roderick Chisholm. In Roderick Chisholm, ed. *Realism and the Background of Phenomenology*. Glencoe, IL: Free Press, 1960, pp. 76–117.

Meneses, Rita W. and Michael Larkin. 2012. Edith Stein and the Contemporary Psychological Study of Empathy. *Journal of Phenomenological Psychology* 43:2, pp. 151–184.

Merleau-Ponty, Maurice. 1945. *Phenomenology of Perception*. Translated by Donald Landes. London: Routledge, 2012.

Merleau-Ponty, Maurice. 1948. *The World of Perception*. Translated by Oliver Davis. London: Routledge, 2004.

Mill, John Stuart. 1872. *An Examination of Sir William Hamilton's Philosophy*. Fifth edition. London: Longmans, Green, Reader, and Dyer, 1878.

Mill, John Stuart. 1891. *A System of Logic: Ratiocinative and Inductive*. London: Longmans, Green, & Co., 1906.

Miller, Izchak. 1984. *Husserl, Perception, and Temporal Awareness*. Cambridge, MA: MIT Press.

Mooney, Timothy. 2007. On the Critiques of Pairing and Appresentation by Merleau-Ponty and Levinas. In C. Cunningham and P. M. Candler, eds. *Transcendence and Phenomenology*. London: SCM Press, pp. 448–494.

Moran, Dermot. 2000. *Introduction to Phenomenology*. London: Routledge.

Moran, Dermot. 2004. The Problem of Empathy: Lipps, Scheler, Husserl and Stein. In Thomas A. Kelly and Phillip W. Rosemann, eds. *Amor Amicitiae: On the Love That Is Friendship. Essays in Medieval Thought and Beyond in Honor of the Rev. Professor James McEvoy*. Leuven: Peeters Publishers, pp. 269–312.

Moro, Valentina, Massimiliano Zampini, and Salvatore M. Aglioti. 2004. Changes in Spatial Position of Hands Modify Tactile Extinction but Not Disownership of Contralesional Hand in Two Right Brain-damaged Patients. *Neurocase* 10:6, pp. 437–443.

Mulligan, Kevin. 1995. Perception. In Barry Smith and David Woodruff Smith, eds. *Cambridge Companion to Husserl*. Cambridge: Cambridge University Press, pp. 168–238.

Noë, Alva. 2004. *Action in Perception*. Cambridge, MA: MIT Press.

Nussbaum, Martha. 2001. *Upheavals of Thought: The Intelligence of Emotions*. Cambridge: Cambridge University Press.

Olafson, Frederick. 1994. Heidegger à la Wittgenstein or 'Coping' with Professor Dreyfus. *Inquiry* 37:1, pp. 45–64.

Peacocke, Christopher. 1983. *Sense and Content: Experience, Thought, and Their Relations*. Oxford: Clarendon Press.

Peacocke, Christopher. 1985. Imagination, Experience and Possibility: A Berkeleian View Defended. In John Foster and Howard Robinson, eds. *Essays on Berkeley: A Tercentenary Celebration*. Oxford: Clarendon Press, pp. 19–35.

Peacocke, Christopher. 1992. *A Study of Concepts*. Cambridge, MA: MIT Press.

Petitot, Jean, Francisco Varela, Bernard Pachoud, and Jean-Michel Roy, eds. 2000. *Naturalizing Phenomenology: Issues in Contemporary Phenomenology and Cognitive Science*. Stanford, CA: Stanford University Press.

Philipse, Herman. 1995. Transcendental Idealism. In Barry Smith and David Woodruff Smith, eds. *Cambridge Companion to Husserl*. Cambridge: Cambridge University Press, pp. 239–322.

Phillips, Ian. 2010. Perceiving Temporal Properties. *European Journal of Philosophy* 18:2, pp. 176–202.

Plato. 2010. *Meno and Phaedo*. Edited by David Sedley, translated by Alex Long. Cambridge: Cambridge University Press.

Priest, Stephen. 2000. *The Subject in Question: Sartre's Critique of Husserl in The Transcendence of the Ego*. London: Routledge.

Putnam, Hilary. 1967. Psychological Predicates. In W. H. Capitan and D. D. Merrill, eds. *Art, Mind, and Religion*. Pittsburgh: Pittsburgh University Press, pp. 37–48.

Putnam, Hilary. 1975. The Meaning of 'Meaning'. In Keith Gunderson, ed. *Language, Mind, and Knowledge: Minnesota Studies in the Philosophy of Science* VII. Minneapolis: University of Minnesota Press, pp. 131–193.

Reid, Thomas. 1786. *Essays on the Intellectual Powers of Man*. Edited by Ronald E. Beanblossom and Keith Lehrer. Indianapolis: Hackett, 1983.

Reinach, Adolf. 1913. The Apriori Foundations of the Civil Law. Translated by John Crosby, in *Aletheia* 3, 1983, pp. 1–142.

Richmond, Sarah. 2011. Magic in Sartre's Early Philosophy. In Jonathan Webber, ed. *Reading Sartre: On Phenomenology and Existentialism*. London: Routledge, pp. 145–160.

Richmond, Sarah. 2014. Inconsistency in Sartre's Analysis of Emotion. *Analysis* 74:4, pp. 612–615.

Ricoeur, Paul. 1967. *Husserl: An Analysis of His Phenomenology*. Translated by Edward Ballard and Lester Embree. Evanston, IL: Northwestern University Press.

Robinson, Howard. 1994. *Perception*. London: Routledge.

Rosenthal, David. 2005. *Consciousness and Mind*. Oxford: Oxford University Press.

Rouse, Joseph. 2000. Coping and its Contrasts. In Mark Wrathall and Jeff Malpas, eds. *Heidegger, Coping, and Cognitive Science: Essays in Honor of Hubert L. Dreyfus, Volume 2*. Cambridge, MA: MIT Press, pp. 7–28.

Russell, Bertrand. 1912. *The Problems of Philosophy*. Oxford: Oxford University Press, 1980.

Ryle, Gilbert. 1949. *The Concept of Mind*. London: Penguin Books, 1990.

Sartre, Jean-Paul. 1936. *The Imagination*. Translated by Kenneth Williford and David Rudrauf. London: Routledge, 2012.

Sartre, Jean-Paul. 1937. *The Transcendence of the Ego*. Translated by Andrew Brown. London: Routledge, 2004.

Sartre, Jean-Paul. 1939a. Intentionality: A Fundamental Idea of Husserl's Phenomenology. Translated by Joseph P. Fell. *Journal of the British Society for Phenomenology* 1, 1970, pp. 4–5.

Sartre, Jean-Paul. 1939b. *Sketch for a Theory of the Emotions*. Translated by Philip Mairet. London: Routledge, 1981.

Sartre, Jean-Paul. 1940. *The Imaginary*. Translated by Jonathan Webber. London: Routledge, 2004.

Sartre, Jean-Paul. 1943. *Being and Nothingness: An Essay on Phenomenological Ontology*. Translated by Hazel E. Barnes. London: Routledge, 1998.

Schear, Joseph, ed. 2013. *Mind, Reason, and Being-in-the-World: The McDowell-Dreyfus Debate*. London: Routledge.

Scheler, Max. 1913/16. *Formalism in Ethics and Non-Formal Ethics of Values*. Translated by Manfred Frings and Roger Funk. Evanston, IL: Northwestern University Press, 1973.

Scheler, Max. 1923. *The Nature of Sympathy*. Translated by Peter Heath. New Brunswick, NJ: Transaction Publishers, 2008.

Schutz, Alfred. 1970. The Problem of Transcendental Intersubjectivity in Husserl. In Alfred Schutz, *Collected Papers III: Studies in Phenomenological Philosophy*. The Hague: Martinus Nijhoff, pp. 51–91.

Schwitzgebel, Eric. 2006. Do Things Look Flat? *Philosophy and Phenomenological Research* 72:3, pp. 589–599.

Searle, John. 1983. *Intentionality: An Essay in the Philosophy of Mind*. Cambridge: Cambridge University Press.

Segal, Gabriel. 2000. *A Slim Book About Narrow Content*. Cambridge, MA: MIT Press.

Sellars, Wilfred. 1956. Empiricism and the Philosophy of Mind. In Herbert Feigel and Michael Scriven, eds. *Minnesota Studies in the Philosophy of Science*, Vol. 1. Minneapolis: University of Minnesota Press, pp. 126–196.

Siegel, Susanna. 2006. Which Properties Are Represented in Perception? In Tamar Gendler and John Hawthorne, eds. *Perceptual Experience*. Oxford: Oxford University Press, pp. 481–503.

Smith, A. D. 2003. *Husserl and the Cartesian Meditations*. London: Routledge.

Smith, Joel. 2010. Seeing Other People. *Philosophy and Phenomenological Research* 81:3, pp. 731–748.

Soffer, Gail. 2003. Revisiting the Myth: Husserl and Sellars on the Given. *Review of Metaphysics* 57:2, pp. 301–337.

Sokolowski, Robert. 2000. *Introduction to Phenomenology*. Cambridge: Cambridge University Press.

Solomon, Robert C. 2003. Emotions, Thoughts and Feelings: What Is a 'Cognitive Theory' of the Emotions and Does it Neglect Affectivity? In Anthony Hatzimoysis, ed. *Philosophy and the Emotions*. Cambridge: Cambridge University Press, pp. 1–18.

Spiegelberg, Herbert. 1976. *The Phenomenological Movement: A Historical Introduction*. The Hague: Martinus Nijhoff.

Stanley, Jason and Timothy Williamson. 2001. Knowing How. *The Journal of Philosophy* 98:8, pp. 411–444.

Stein, Edith. 1917. *On the Problem of Empathy*. Translated by Waltrout Stein. The Hague: Martinus Nijhoff, 1970.

Talero, Maria. 2005. Perception, Normativity, and Selfhood in Merleau-Ponty: The Spatial 'Level' and Existential Space. *Southern Journal of Philosophy* 43:3, pp. 443–461.

Thomasson, Amie. 2007. Conceptual Analysis in Phenomenology and Ordinary Language Philosophy. In Michael Beaney, ed. *The Analytic Turn: Analysis in Early Analytic Philosophy and Phenomenology*. London: Routledge, pp. 270–284.

Tye, Michael. 2002. Representationalism and the Transparency of Experience. *Noûs* 36:1, pp. 137–151.

Tye, Michael. 2003. *Consciousness and Persons: Unity and Identity*. Cambridge, MA: MIT Press.

Williford, Kenneth. 2011. Pre-Reflective Self-Consciousness and the Autobiographical Ego. In J. Webber, ed. *Reading Sartre: On Phenomenology and Existentialism*. London: Routledge, pp. 195–210.

Williford, Kenneth. 2013. Husserl's Hyletic Data and Phenomenal Consciousness. *Phenomenology and the Cognitive Sciences* 12:3, pp. 501–519.

Wittgenstein, Ludwig. 1953. *Philosophical Investigations*. Translated by G. E. M. Anscombe. Oxford: Blackwell, 1998.

Woodruff Smith, David. 2007. *Husserl*. London: Routledge.

Wrathall, Mark. 1999. Social Constraints on Conversational Content: Heidegger on *Rede* and *Gerede*. *Philosophical Topics* 27:2, pp. 25–46.

Zahavi, Dan. 2003. *Husserl's Phenomenology*. Stanford, CA: Stanford University Press.

Zahavi, Dan. 2005. *Subjectivity and Selfhood: Investigating the First-Person Perspective*. Cambridge, MA: MIT Press.

Index